Spotlight on the UK

Spotlight on the UK

P. M. Turner

General Editor: Michael Naish

M

Macmillan Education

First published 1978

Published by
MACMILLAN EDUCATION LIMITED
Houndmills Basingstoke Hampshire RG 21 2XS
and London
Associated companies in Delhi Dublin
Hong Kong Johannesburg Lagos Melbourne
New York Singapore and Tokyo

Filmset by BAS Printers Limited,
Over Wallop, Hampshire
Printed in Hong Kong

British Library Cataloguing in Publication Data
Turner, P M
Spotlight on the UK. – (The world in focus).
1. Great Britain – Description and travel – 1971 –
I. Title II. Series
914.1 DA632

ISBN 0-333-14366-3

Contents

Preface

The World in Focus

This series of four books has been designed for the student preparing for examinations at sixteen-plus. These books have been constructed to provide comprehensive coverage and a flexibility of approach which is relevant to all modes of examination and a wide variety of teaching techniques.

The authors have set out to introduce students to themes of continuing importance in the world of today. These themes, problems and concepts are studied in various national or regional settings, and it is hoped that students will be encouraged to cross-refer from one volume to another in the study of a particular theme.

In this volume, P. M. Turner asks important questions about what is happening in the United Kingdom. He involves the reader in an examination of the processes which influence the location and distribution of phenomena which affect our daily lives. Since decisions about the spatial arrangement of the landscape are commonly controversial, he also asks us to consider the conflicts which arise about such decisions. The book is intended to involve the reader in the process of inquiry. Consequently a wide range of resources is presented and the reader is invited to analyse the data they provide in order to investigate the questions and problems which are raised. This presents the opportunity for discussion, for simulation and for the practice of a variety of skills. There are suggestions for further work, including project and fieldwork, and the enthusiastic reader will find great stimulus for such work in this book.

M. C. Naish

For the Student

This book contains many questions. It is important to realise that there is not always a right answer. Very often, the answer is a matter of opinion. You will find in chapter 1 that people see the world differently. When there are problems, different people think of different solutions. There are conflicts to which there are usually no easy answers. The best answer you can give is one which uses any evidence you have to support your opinion.

An important reason for reading this book is to find out what problems there are in the UK and what conflicts grow from them. Another purpose is to help you think out your own opinion by understanding the way things happen and the way they change. If you can do this, you can help solve problems, both now and when you leave school.

If you are to take your study further you will need to find additional information for yourself and you may have the opportunity to make use of fieldwork, the library, or any other sources of information that you can find. You will be making the best use of this book if you try to be really thorough in your study, following up the topics and ideas which you find in these pages.

P. M. Turner

Acknowledgements

The author and publishers wish to acknowledge the following photograph sources:

Aerofilms Fig. 5. Peter Baker Fig. 64f. Barnabys Picture Library Fig. 82. Godfrey Bell Fig. 227. British Airports Authority Fig. 205. British Airways Fig. 207. British Insulated Callenders Cables Ltd Fig. 225. British Petroleum Fig. 161. British Railways Board Figs. 192, 193. British Steel Corporation Fig. 16. British Tourist Authority Figs 80, 121a (ii). British Transport Docks Board Figs. 197, 199. J. Brownbill Figs. 64a, 64e, 65, 108b, 113. John Burrows Fig. 145. Camera Press Figs. 31a, 31b, 31c, 31d, 31e, 31g, 193. Central Electricity Generating Board Fig. 41. Central Press Photos Figs 105, 117, 181. Civic Trust Library Figs 1, 2a, 2b, 2c, 43, 180. Freddie Collins Fig. 81. Coloursport Fig. 64a. Cumbernauld Development Fig. 137. Daily Express Fig. 211. Department of Leisure Services, Torbay, Fig. 80d. John Dewar Studios Figs 158, 160. Esso Photograph Fig. 78. Ford of Britain Fig. 46. Gazette Series and West London Observer Fig. 182b. Graham Photography Fig. 64c. Greater London Council Fig. 121 b (ii). The *Guardian* Fig. 164. Archie Handford Ltd Fig. 39. Imperial Chemical Industries Ltd Fig. 30. Industrial Research Building, Birmingham Figs 142b, 143a, 143b. R. Jones, Liverpool Council Fig. 64b. Martin Koretz Fig. 146. London Express Pictures Fig. 188. Mansell Collection Figs 4, 15. National Coal Board Fig. 33. NFU Photo Service Figs 221, 223a, 223b, 223c, 223d, 224, 226a, 226b, 229. Pacemaker Press Fig. 112a. Peak Park Planning Board Fig. 67. Press Association Figs 112b, 112c. Terence Soames Fig. 18. Scottish Tourist Board Fig. 154. Sport and General Press Agency Fig. 219. *Sunday Times* Figs 38a, 38b, 213. By courtesy of Swan Hunter (Shipbuilders) Ltd Fig. 24. Syndication International Figs 31f, 80a, 80b, 80c, 80f, 80g, 80h, 209. John Topham Figs 84, 218.

The author and publishers wish to thank the following who have kindly given permission for the use of copyright material:

Edward Arnold Ltd for a map from *A Study of Urban Geography* (1974) by Carter and Davies. British Railways Board for extract from British Rail Sealink Pamphlet. Caledonian MacBrayne Ltd for extracts from their Passenger Service Timetable. John Carvel for extracts from 'Rich-poor gap stays', from the *Guardian* 28 December 1973. *Daily Express* for the article 'Honeymooners Hit in Airport Chaos' from the *Daily Express* of 2 September 1974. The Economist Newspaper for the article 'Scotland's Oil' from *The Economist* of 20 January 1973. The Geographical Association for Figure 1 from 'Milford Haven and its oil industry 1958/1969' by D. G. Watts in *Geography* Vol. 55, 1970, reprinted by permission of the author and publishers. The Geographical Magazine for two maps from 'Resources for Britain's Future' by J. T. Coppock in *Geographical Magazine* February, 1970. Mike Harding for 'Hattersley Lament'. Heathrow Association for the Control of Aircraft Noise for permission to reproduce the HACAN poster. HMSO for two tables from 1971 Census Reports and a map of Pembrokeshire National Park in the Pembrokeshire Coast leaflet based upon the Ordnance Survey Map and reprinted by permission of the Controller of Her Majesty's Stationery Office. Hutchinson Publishing Group Ltd for a map from *The Geography of Energy* by Gerald Manners. The Institute of British Geographers for Figures 4 and 10 from an article entitled 'The Distribution of New Commonwealth Immigrants in the London Borough of Ealing, 1961–66' in *Transactions* IBG, Volume 58, March 1973. E. Jones for illustration from *Social Geography of Belfast* (1960). Longman Group Ltd for two maps from *A Study of Urban Geography* (1974) by Carter & Davies. Thomas Nelson & Sons Ltd for a map from *The British Isles: A Systematic Geography* edited by J. Wreford Watson and J. B. Sissons (1964). Oxford University Press for Geology Map from the *Concise Oxford Atlas* by P. D. Bickmore. Peak Park Joint Planning Board for map from Goyt Valley Traffic Scheme leaflet. The Queens University of Belfast for diagram from *Irish Geographical Studies* by F. W. Boal. Times Newspapers Ltd for material from *The Sunday Times*, 'Thomson's Poor Regions' and 'Brandt's Poor Regions' (23 December 1973), 'Routes to Europe and East' (8 July 1973) and 'Nigg Bay Cromarty' (7 October 1973). Transworld Feature Syndicate (UK) Ltd for 'The Rape of Manchester' from *Observer Magazine*. Copyright © Observer Magazine. West London Post for extract 'Traffic Stops for M3 Demos' from the *Shepherds Bush Gazette & Hammersmith Post*, 11 July 1974, by permission of the editor. Which? Consumers' Association for diagrams from April 1973 issue of *Motoring Which?* John Wiley & Sons Ltd for Figure 1B, Railway freightliner routes and terminals, from *Regional Development in Britain* by Manners, Keeble, Rodgers and Warren, and a diagram from *Europe 1952* by S. van Valkensburg and C. C. Held, published by John Wiley & Sons Inc. New York.

The publishers have made every effort to trace the copyright-holders but if they have inadvertently overlooked any, they will be pleased to make the necessary arrangement at the first opportunity.

1 Things are not what they seem

Figure 1

Figure 2 (a) (above left) (b) (above right) (c) (below)

What is beauty?

Do you think this street is ugly?

If you do, what is it in particular that makes you think so? Perhaps you do not consider this to be an ugly street, but many people do think that the great clutter of road-signs spoils it. If this is the case, why were the signs put up in the first place? It is probably because the people who decided the signs should go up were thinking more about road safety than beauty. Then, surely, they were right. There are far too many road accidents as it is, and anything that can be done to reduce them must be worthwhile. But, equally, you might say that anything beautiful is worthwhile and should not be defaced.

Then who is right? Well . . . it depends on who you are; on how you 'see' your surroundings. There is really no 'right' answer to the problem. Here are a few more pictures to make you think.

1 In picture (a), which building do you think is out of place? Assume that one of them is.

2 Do you think the motorway in picture (b) is beautiful?

3 In picture (c), which building do you think is the wrong size to fit in with the others?

When you have answered the questions, check with others in your group to see if they agree with you. If not, try to give each other reasons for your opinions.

We shall see many times, later in this book, how people disagree about what should be done in different parts of Britain, but for the time being, let us concentrate on the idea that people 'see' things in different ways.

When is a resource not a resource?

4 Do you know what is meant by the word 'resource'? If not, look it up in a dictionary.

Some resources have not always been regarded as such. This is not just a question of people seeing things differently. As time passes, new inventions are made which change the way people regard the land. The following maps may help to show what is meant.

5 Look at map 3 (b). Where is the highest density of population outside London?

6 Look at map 3 (c). Where is the lowest density shown?

7 Compare maps (a) and (b). Are high densities of population found on or near the coalfields?

8 Now compare maps (a) and (c). Do the areas of high population density appear to be in roughly the same areas as the coalfields?

9 What is meant by 'high' population density? Were there *any* areas of high density in 1700?

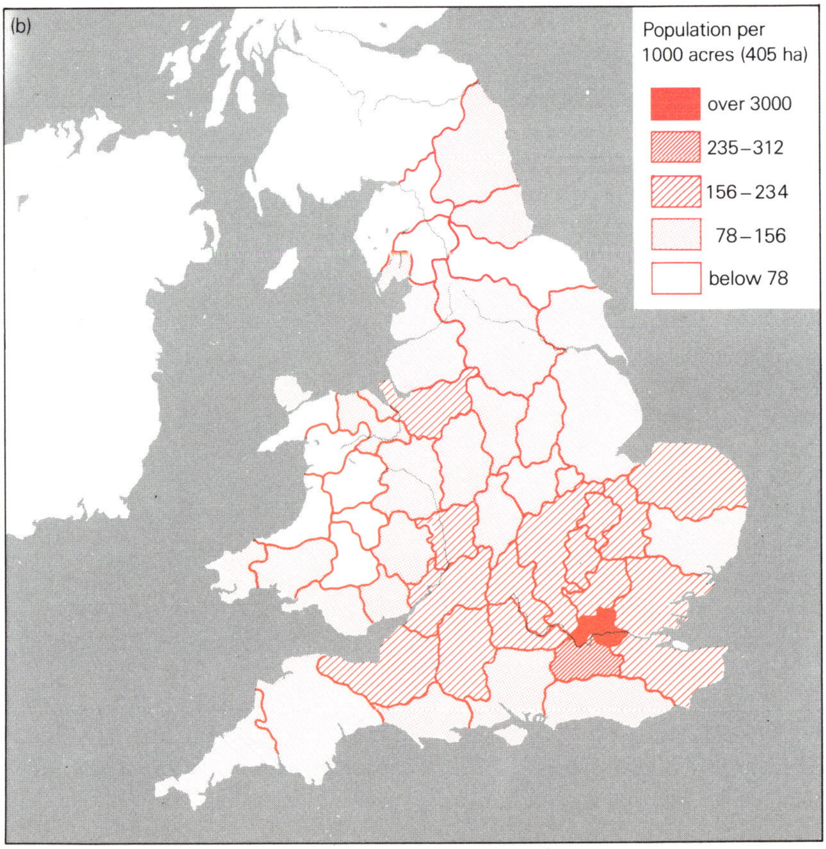

Figure 3 (a) Coalfields of Britain
(b) Approximate density of population in 1700
(c) (opposite above) Density of population in 1911

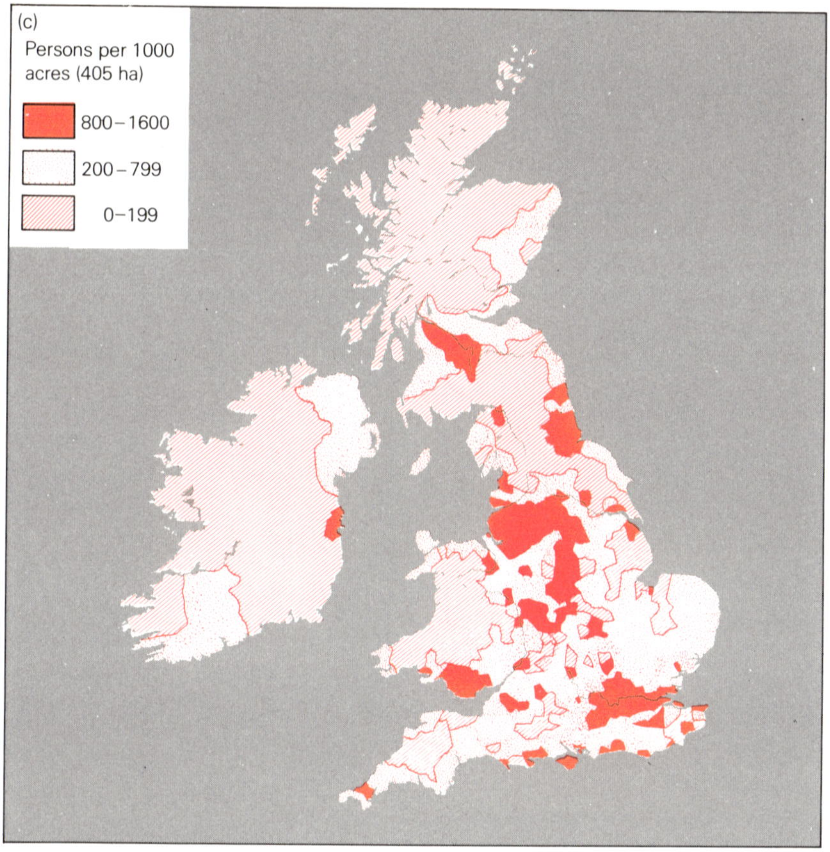

Figure 4 Manchester in the nineteenth century

Before steam power was invented, no one bothered much about coal. Some of the rich people in London used it for burning in their homes, but most used wood. The coal was there all the time. People even knew it was there, but they had no great use for it. After the invention of steam power and the use of coke for smelting iron ore to make iron and steel, the picture was much different. Great factories were built on the coalfields and there was a huge movement of people into these areas to work.

Coal was now a resource, but it too brought its problems.

10 Do you think Manchester was beautiful in the nineteenth century?

Perhaps you do not 'see' Manchester as very attractive in figure 4, but it is quite possible that a mill owner of the time would have thought it the finest sight on earth.

11 Why might a mill owner have had such an opinion?

The example of coke and coal is a fairly obvious and well-known one. Did you know that the area known as the Fens, which is now one of the richest areas of farmland in Britain, was once marsh? Until the sixteenth and seventeenth centuries, when ways of draining the marshes were discovered, this area was all but useless – a hideout for robbers and outlaws. If you did not know this, read the story of Hereward the Wake who was a kind of Robin Hood of the Fens.

Now look at the air photograph of the fen country in figure 5, and, if possible, at the OS 1:50000 sheet 143.

12 What do you notice immediately about the pattern formed by the rivers and small streams? The minor roads follow the same pattern.

13 You should now be able to guess the simple way in which most of the drainage was done. Explain how it was carried out.

It is difficult to imagine that this open farmland was once so treacherous that if you did not know the path, or have a local guide, you were likely to step off into seemingly bottomless mud. Perhaps even more surprising is that the mountain areas of Britain, the Lake District, Snowdonia and the Scottish Highlands, were once thought of as dreadful places, to be avoided at all costs. Gradually, in the eighteenth and

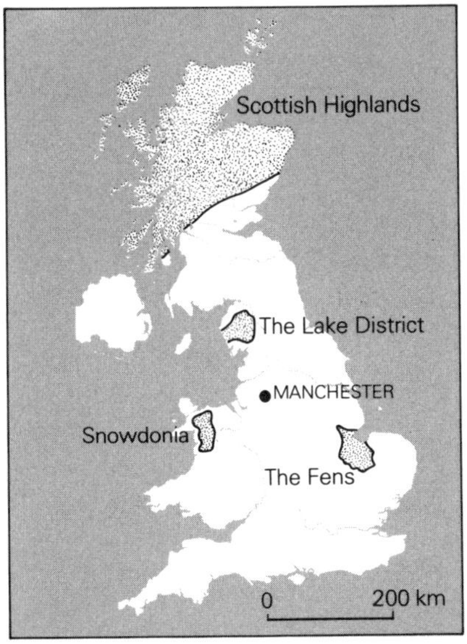

Figure 5 The Fens today
Figure 6 (above right)

nineteenth centuries, opinions changed. People began to think of mountains as romantic and noble. William Wordsworth, who lived in the Lake District, wrote:

'Ye presences of Nature in the sky
And on the earth! Ye visions of the hills!
And Souls of Lonely places!'

Poets like Wordsworth helped to change the image of the mountain areas. Since Wordsworth's time the popularity of the highlands has increased steadily so that now people flock to them at holiday times. Ordinary men and women, who two centuries ago would have found the rocks and fells repulsive, now consider them to be among the most beautiful places.

Images

In the last paragraph, the phrase 'change the image' was used. It was used deliberately, because often people have an 'image' of a place which may not be the same as someone else's. It often depends on how much they know about it, or perhaps on fashion. You could say, for example, that at one time mountains were not in fashion, but that now they are, and that people's image of them has changed from being frightening and horrible to being beautiful. It is quite likely that your mental image of a part of Britain which you do not know well is quite different from the image of it of someone who lives there. Neither of you will know everything about the place, though you may both find out more as time goes on.

Now look at the maps in figure 7. The dark shaded areas are the ones which senior pupils at four schools thought were the best places in Britain to live. The place containing the school is named on each map.

14 Name a part of Britain which all the pupils like.
15 Do all the children like their home areas?
16 Are any of the popular areas well-known for holidays?
17 If you had a choice, where in Britain would you most like to live? What are your reasons?

Twenty schools took part in the survey for the maps but there is not space to show them all. However, they all liked their home areas and they all tended to like the south of England, though less so if they lived

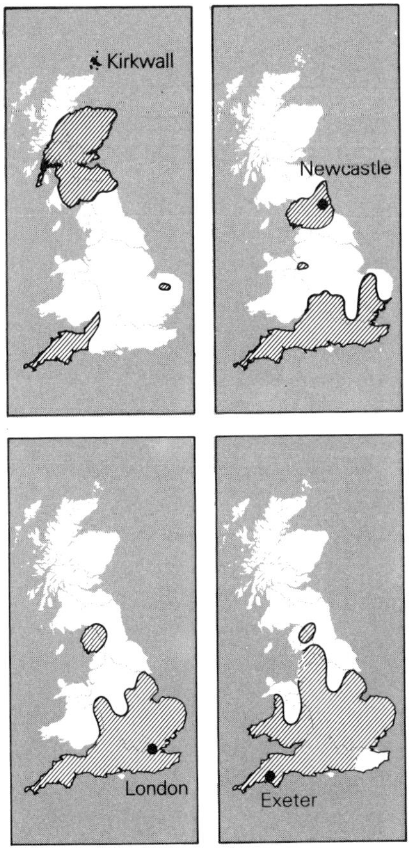

Figure 7 Regional images

further away.

18 Why do you think so many people prefer the south of England?
19 What do you mean by 'the south of England'? Name the coun-

Figure 8

ties contained in it. (Use an atlas to help you.)

20 Now name the counties contained in the north of England.

21 Draw a map to divide England into north and south.

Many people have quite definite ideas as to what is 'the North' or 'the South'. These ideas are probably quite different according to where the people live. Look at the photographs in figure 8 of roadsigns on the M1 and M18. Look up the places named in an atlas. Now where is the North?

There are also some traditional views about what the North and South are like. Northerners are supposed to think that southerners are 'soft' and that neighbours are 'unfriendly'. If you are a southerner, do you see the North in the way that Doncaster council think you might?

We have seen that different individuals have different opinions about places. One thing (but only one of several) which affects their view is how well they know the place. It is perhaps obvious to say that the more familiar one is with a place, the more accurate one's knowledge of it. But can anyone ever know any place perfectly? For instance, how well do you know your home (or your nearest) town?

22 How far is it from your home to the town centre?
23 How far is it from your home to school?
24 How far is it from your school to the town centre?
25 How many different bus routes pass through your town centre? (If you live in a big city, see if you know how many bus routes pass through your nearest suburban centre.)
26 Where is the town centre? To answer this question, imagine you are explaining where it is to a stranger.

The last two questions show more ways in which views may differ. For instance, how big is a 'large' city? How big does a shopping centre have to be before it is called a suburban centre? And, of course what *is* a town or city centre? You can look up the answers to questions 22 to 25 quite easily, but for question 26 check with others in your group to see if you agree about where the town centre is. If you disagree with someone, give each other reasons for your choice.

What is the UK?

At this point you can consider how much you know of this country. So far in this chapter the words 'Britain' and 'England' have been used, although the book is about the UK.

In an encyclopedia or other reference book, find out the answers to the following questions:

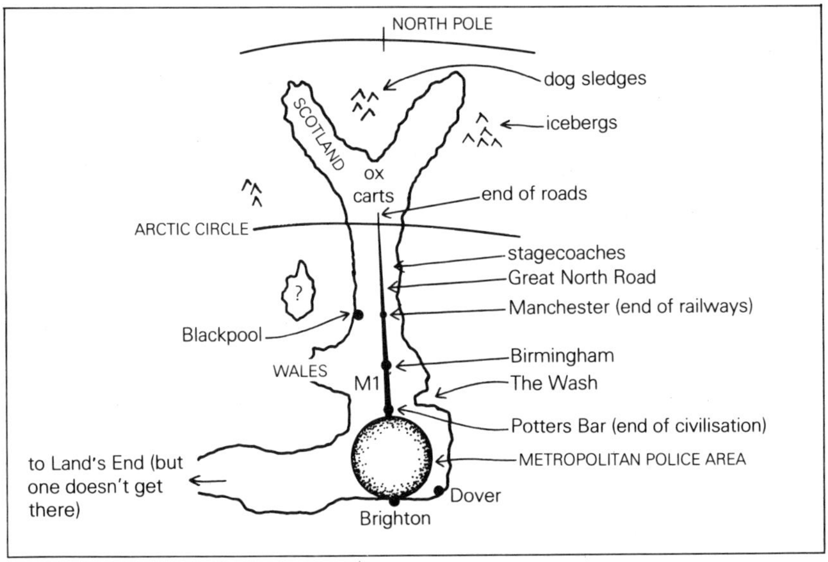

Figure 9 How Londoners see the North – at least according to the Doncaster and District Development Council

27 What is meant by the letters UK?
28 Which countries are included in it?
29 What is meant by the letters GB?
30 Which countries are included in it?
31 What land area and which countries are included in 'the British Isles'?

In fact 'Britain' is a vague term. This does not always matter, but if you wish to be precise you should use the names of the countries or one of the terms in the questions above.

Having decided what is meant by the UK we should consider its place in the world. For instance, is the UK part of Europe? Well, it is part of the same land mass, though separated by a shallow sea, but we do not speak the same language as any European country. Is the UK part of the Commonwealth? Yes it is, but some Commonwealth countries feel closer ties with their neighbours than with distant Britain. Then what about the USA? It is a foreign country but they speak more or less the same language. Much of our trade is with the USA and we rely on the Americans for defence, despite having the Atlantic Ocean between us.

32 Trace the countries of Europe from an atlas. Find out which of them belong to the European Economic Community (better known as the Common Market) and shade these on the map with horizontal lines.
33 Find out which European countries belong to the North Atlantic Treaty Organisation (NATO) and shade them on the same map with vertical lines.
34 Finally, shade in the European countries in the Commonwealth with diagonal lines.

Of course, the UK is part of the Commonwealth and of Europe and is closely tied to the USA, but politicians disagree strongly about how close our connections should be. Many consider that we should not be in the Common Market. Others think we should loosen our ties with the USA. Some say that the Commonwealth is out of date and not of great value in the modern world.

There are good reasons for all these opinions and among them, as we have seen in this chapter, will be the different ways in which individuals 'see' the role of Britain.

Understanding Britain

Finally, it may seem strange to start a book on the UK by talking about the way people see things, or about their images of places, but this is most important. Almost all the landscape of Britain, rural as well as urban, is man-made. Someone, or some group of people, **decided** that something should happen (a field should be ploughed, or a motorway be built – there was a decision). Now how can you begin to understand how decisions are made, and why others disagree with them, unless you know that people see things differently? Decisions result in changes (look back to the preface to see why these are important).

Once you begin to understand the way things work, you may feel that you want to help to decide some local, or even national, issue, either now or when you have left school. If you feel that you can participate in local decision-making, then so much the better. If you can understand local problems outside your own area, and national ones, then better still.

PROJECTS

1 As you might guess from the chapter you have just read, this project is intended to illustrate how people see and remember things differently. So, when you first do it, *make sure that you do not look up any answers or ask anyone else.* Remember, there is no 'right' and no 'wrong', just your own opinions and memory.

Try to draw a map of your home town, or nearest town, from memory. Use a large sheet of squared paper and try to keep to an approximate scale. Do not draw a detailed street plan. Instead, try to include the following items:

(a) The boundaries of the neighbourhood in which you live;
(b) The major roads in the town centre;
(c) Major landmarks, like important buildings, squares, statues etc., in the town centre.
(d) Name, and show by shaded areas, neighbourhoods and suburbs other than your own which you know by name. Do *not* try to show the boundaries of these areas accurately.
(e) Name, and show as a shaded area, where you would most like to live in the town. If it is one of the neighbourhoods already shown, underline the name.
(f) Do the same for the area where you would least like to live.

2 Small group project:
(a) Display everyone's memory map in the classroom;
(b) Draw a master map as follows:
(i) mark in red the items such as streets, landmarks, suburbs etc. which everyone has included;
(ii) mark in blue items shown by over three-quarters of the class;
(iii) mark in yellow the places shown by less than a quarter of the class;
(iv) draw a thick red arrow pointing to the place where most people would like to live;
(v) draw a thick grey arrow pointing to the place where most people would least like to live;
(c) Discuss the reasons for the choice of popular and unpopular places to live. Try to decide for yourself whether the reasons are 'real' or whether they result from not knowing enough about the areas. Were people's popular choices near their homes? Were the unpopular places a long way away?

3 Small group project:
(a) Try to write, from memory, a visitors' handbook to your town. It should include:
(i) the names of the main hotels;
(ii) the names of the chief places of entertainment, sport and parks;
(iii) the names of the main museums, art galleries and places of interest;
(iv) information about any special attractions of the town, or items which are particularly interesting.
(b) Try to draw a group memory map of the town centre, such as you would expect to find in a visitors' handbook, showing the head Post Office, main line railway station, town or city hall and the names of the main shopping streets.
(c) After completing parts (a) and (b), look at the handbook for visitors produced by your town (most towns have one). Then:
(i) make a list of all those places under headings (i) to (iv) in part (a) which are included in the official book but not in yours;
(ii) discuss with your group the reasons why they were left out. Was it because:
(a) you did not know they existed?
(b) they were on the far side of the town from where you live or go to school?
(c) you were not interested in those particular items?
Of course, there may be other reasons, these are just to start you off.
(iii) make a list of all the places which you have included but which are not in the official book.
(iv) again, discuss the reasons for the difference.
(d) Do you think the official handbook would be better if it left out some items and included others? Make a list of your suggested

changes, explaining your reasons very carefully.

(e) Some towns do not have an information handbook. If yours is one of these, you may consider that it ought to have one. If you think there should be such a booklet, write a polite letter to the town clerk's office suggesting it. Be sure to explain carefully your reasons. Your teacher may decide that one of the letters can be posted.

2 Make it here

Location

Imagine you are a miller in medieval Britain, wishing to set up a new mill. You will have to buy corn from the villages and you want to sell flour to the bakery in the town. You will have to use water power and you know that where the river narrows is a good place for a mill pond and water wheel.

In reaching your decision about where to place the mill you will have to take into account different costs if you are to make as much profit as possible. The main costs would be:

(a) the price of corn.
(b) the price per kilometre charged by the packman for carrying the corn.
(c) the cost of carrying out the milling.
(d) the price per kilometre charged by the packman for delivering the flour to the bakery.
(e) the price paid by the bakery for the flour.
(f) the cost of building a mill race, which would be greater where the river was not naturally suitable.

You cannot do much about items (a), (c) and (e). You can assume they are fixed. However, to reduce your costs you can choose a narrow stretch of river where the distance travelled by the two packmen is as short as possible.

Which place on the map in figure 10 would you choose?

In solving this problem you have been working on the **location** of the mill.

The word 'location' means *where* something is. But 'where' does not just mean, say, the latitude and longitude. It includes, 'where in relation to other things connected with it'. As well as *describing* where something is, it is also important to *explain* why it is there. What reasons can you give for the choice of location for your medieval mill?

The reason for choosing a particular location was to make as much profit as possible. To put this another way, you could say that the place was chosen to cut down costs as much as possible.

Having looked at an imaginary location problem, you can now try your skill on a real one. Look at figure 11 and decide where you would put an ironworks. Do *not* look at figure 12 yet!

In 1851 it needed several tonnes of coal to make one tonne of iron. Both coal and iron are very bulky and heavy, so they are expensive to transport. In view of the transport costs involved, it should not be very difficult for you to say where the ironworks would be found. When you have made your decision, look at figure 12 to see where they actually were.

Figure 10 Imaginary sites for a mill

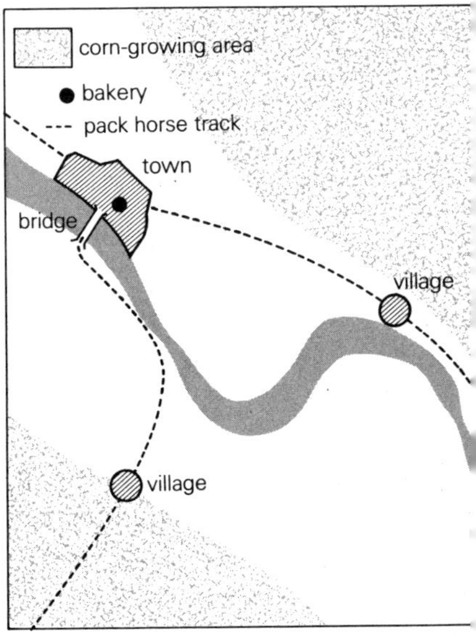

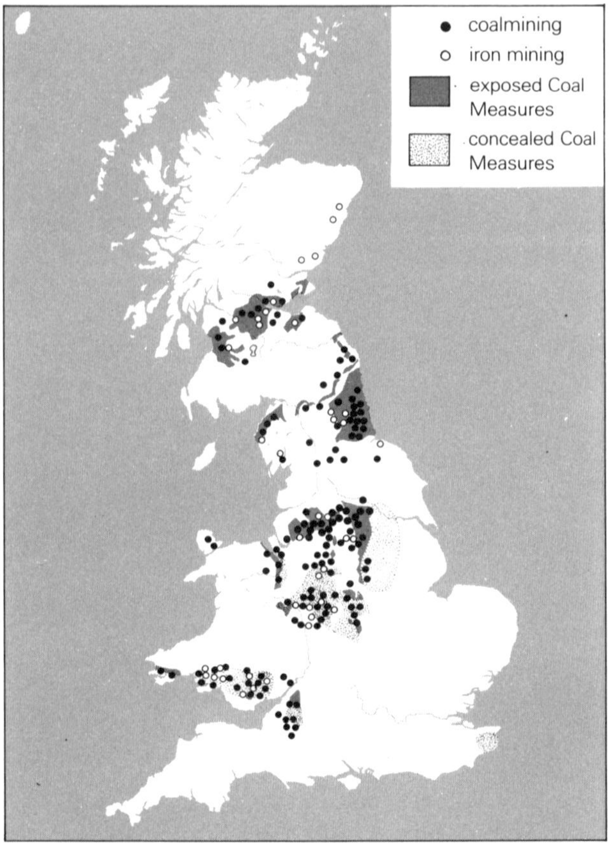

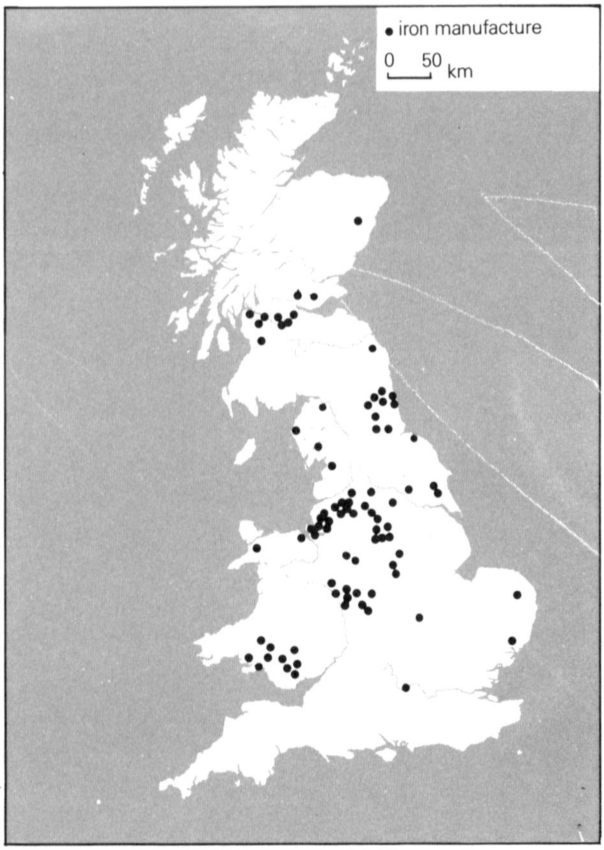

Figure 11 (left) Location of coal and iron mining in 1851
Figure 12 (right) Location of iron manufacture in 1851

Since medieval times, and particularly since the industrial age began, location has become much more complicated. Among other things, transport costs are not nearly so important as they were, though they still apply to some industries, as we shall see.

Processes in locating industry

One of the processes involved in locating industries might be thought of as a chain reaction. The presence of one industry attracts another. The simplest kind of chain is where the finished product of one industry is the raw material of the next. For instance, steel slabs may be one of the finished products of a heavy steelworks; these slabs are then the raw material of a special steelworks; the special steel then becomes the raw material of a magnet-making works; then the magnets become one of the raw materials of a hi-fi-producing factory. Another similar kind of chain is found when industries which make the same sort of things are attracted to the same areas. They would be able to use the same suppliers of raw materials and they may be able to sell to similar buyers. For example, where there is shipbuilding, firms making marine engines, pumps and heavy gears might be attracted, particularly if there are steelworks in the area to supply them with raw materials. Over a period of time, firms making similar products might build up a pool of skilled labour in their area and this in turn might attract more of the same type.

Growth of factories

Another process that started about the middle of the eighteenth century was the increasing use of machinery. The invention of the steam-engine, the railway, and several different textile-making machines was particularly important. These inventions took place at a time of great industrial growth, sometimes called the Industrial Revolution. These new machines were best used in factories, because the more machinery you have together in one place and the more people working on it, the more efficient will be production. For example, a big cotton-making machine does not cost much more than a small one, but it does produce a *lot* more thread. So if you can afford a bigger machine, you make more profit. This is known as making use of **economy of scale**.

1 Explain in your own words the

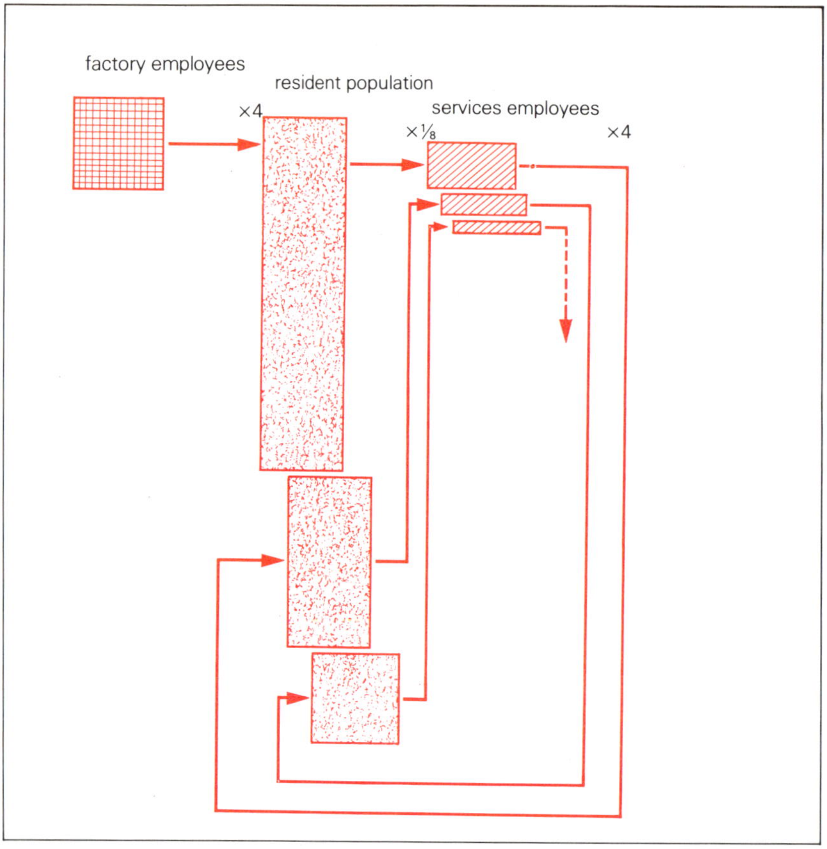

Figure 13 Town development diagram

meaning of 'economy of scale'.

The use of coal for smelting iron and for steam power, together with industrial chain reactions and economies of scale, meant that huge concentrations of industry grew up on the coalfields.

Growth of conurbations

There is another type of chain reaction to consider.

Here is an explanation of the diagram in figure 13.

Imagine there is a new factory which employs a hundred men.

Each man is married with two children. How many people is that altogether?

All these people need shops, churches, schools, etc. (known altogether as 'services').

Every 400 residents in an area need fifty people working to provide services. These fifty people are all married with two children. How many people are there now in the area?

2 Explain in your own words the meaning of the signs, $\times 4$ and $\times \frac{1}{8}$ on figure 13.

Now that you can see how this system works, imagine what was happening during the Industrial Revolution. There was not just one small works of one hundred workers, but an ever-increasing number of bigger and bigger factories. The places where the services and the houses of the workers are to be found are the great industrial towns. On the coalfields, the older towns (which were road junctions and which rapidly became rail junctions) acted as the main growth points. The smaller towns and villages also grew into cities and towns. Eventually

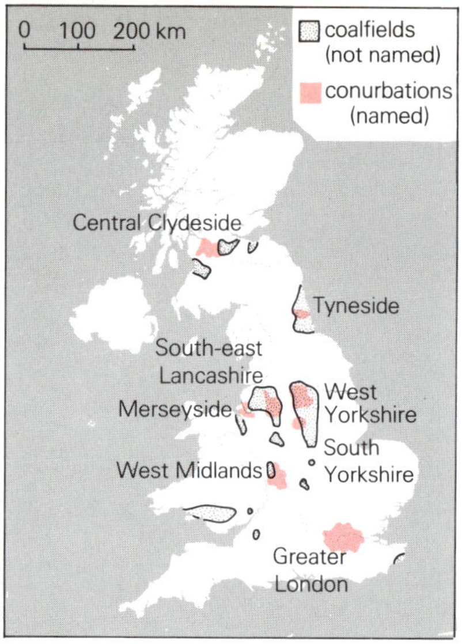

Figure 14 Conurbations

many of them expanded so much that their built-up areas merged. These huge areas of settlement are known today as **conurbations.**

3 How many conurbations are there in the UK?

4 How many of them are on, or partly on, a coalfield?

This industrial and urban growth made Britain the most prosperous country in the world, but there were also drawbacks. The rush of people to the cities and the huge growth of population led to overcrowding. In the early days, working conditions in the coal mines and factories were intolerable. Pollution and disease were widespread. At first people did not realise what they were doing to the countryside, but even as early as 1854 C. F. Cliffe wrote of the Swansea Valley:

'. . . those not bred to the work from childhood cannot stand the sulphurous atmosphere and heat. The copper smoke is a serious nuisance to the country around . . . and has on several occasions afforded employment to the gentlemen of the long robe. At night it was no bad representation of the infernal regions, for the smell aids the eye. Large groups of

Figure 15 Wolverhampton in 1866

odd chimneys and rickety flues emit sulphurous arsenical smoke or pure flames; a dense canopy overhangs the scene for several miles, rendered more horrible by a peculiar lurid glare.'

5 Try to find out where the Swansea Valley is.

6 From C. F. Cliffe's description, what sort of industries went on there?

7 Who do you think are the 'gentlemen of the long robe'?

8 Find out what kinds of industry are found today in the Swansea Valley.

9 Look up the location of Wolverhampton in an atlas.

10 Imagine you are a newspaper reporter in 1866 looking at the view shown in figure 15. You wish to give your readers a vivid impression of what it is like. Remember, in those days there was no television, no films and very few pictures of any kind.

Bearing all this in mind, try to write a short description of the scene for an imaginary newspaper. In your description, see if you can use any of the words which C. F. Cliffe wrote.

The steel industry

This is Sidney Johnson. He is working on a blast furnace at a small steelworks in Sheffield. He works in this factory partly because the work is varied, because the pay is fair and because the hours are reasonable, even though there is shift work. He works in steel because the rest of his family are steelworkers and because his father was also.

How important is the steel industry in Sheffield today? The following table (figure 17) will help. (Tables of figures sometimes put people off. However, they are only provided to help people find out information. Do not try to look at *all* the numbers. A good way of looking is to *compare* a few different numbers with each other. One way of doing this is to work out the numbers as a *percentage* of the total. If you do not know how to do percentages, ask someone to help you.)

Figure 16 Sheffield steelworker

Figure 17 Employment in Sheffield, 1971. Source: Department of Employment

Primary industries		
I	Agriculture, forestry, fishing	267
II	Mining and quarrying	2840
Manufacturing industries		
III	Food, drink and tobacco	9413
IV	Coal and petroleum products	(Included elsewhere)
V	Chemicals and allied industries	316
VI	Metal manufacture	39 123
VII	Mechanical engineering	12 136
VIII	Instrument engineering	1222
IX	Electrical engineering	2781
X	Shipbuilding and marine engineering	6
XI	Vehicles	5005
XII	Metal goods not elsewhere specified	44 511
XIII	Textiles	154
XIV	Leather, leather goods and fur	174
XV	Clothing and footwear	1035
XVI	Bricks, pottery, glass, cement, etc.	5015
XVII	Timber, furniture, etc.	3090
XVIII	Paper, printing and publishing	2659
XIX	Other manufacturing industries	1860
Construction industries		
XX	Construction	15 398
Service industries		
XXI	Gas, electricity and water	3315
XXII	Transport and communications	11 258
XXIII	Distributive trades	30 792
XXIV	Insurance, banking, finance and business services	6350
XXV	Professional and scientific services	36 323
XXVI	Miscellaneous services	14 581
XXVII	Public administration and defence	7698
Unclassified		4759
TOTAL		262 081

NOTE:

1 This table relates to the Sheffield, Attercliffe and Woodhouse Employment Exchange Areas

2 The figures are based on a count of National Insurance Cards and show total employees (employed and unemployed) for 1971

11 What percentage of all workers are involved in metal manufacture? (Item VI).

12 What kinds of products might be included under the heading, 'Metal goods not elsewhere specified'? (Item XII)

13 What percentage of the total is engaged in this work?

Steelworkers would be included in the first of these two categories and cutlery-making in the second. If you add the two together you will see that steel and steel goods are still very important in Sheffield. Even so, look how many people are employed in the various jobs called 'service industries'.

14 What percentage of the total is employed in service industries?

You can see that even in Sheffield, less than half the people work in steel while more than half work in services. Look back to figure 13 and its explanation to see how the services of a town come to grow.

Although Sheffield is famous for steel, it does not produce very much compared with other parts of Britain. This is because Sheffield's fame comes from making high-quality steel for special purposes. The biggest steelworks in Britain produce *crude* steel, for making into girders, or they produce sheets, strips and slabs for general purposes.

All steelworks employ a great many workers, but at the largest there are many thousands. Partly because so many jobs depend on it and partly because steel is used in so many things, steelmaking is one of the most important industries in Britain.

Integrated steelworks

Look at the picture of the Margam and Abbey works at Port Talbot in South Wales shown in figure 18. The biggest steelworks nowadays make

Figure 18 The Margam works, Port Talbot

both iron and steel. The iron is made in blast furnaces and while it is still molten it is taken to the steelmaking area, which is a different part of the same works. While the steel is still hot it is rolled into slabs, or sheets or strips.

Location of the steel industry

The harbour in the foreground of the photograph in figure 18 is a new one, specially built for the steelworks.

15 Why do you think a harbour is important? If you cannot decide for yourself the table on the right might help.

A table of figures showing different years is helpful to show **trends** that are occurring over a period of time.

16 What has happened to the consumption of home ores during the five years shown on the table?

17 What has happened to the consumption of imported ores during the same time?

Now perhaps the reason for a new harbour is obvious.

Look at figure 20 and back to figure 11. Shortly after 1851 the iron ore in the coalfields ran out. New iron ore was found in the other places shown in figure 20. The new ore was *low-grade*. This means that there was not much iron in it and therefore several tonnes would be needed to produce one tonne of pure iron. But although more iron ore was needed, new inventions meant that less and less coal was required to make one tonne of iron.

18 If you had to build a new steelworks under these conditions, where would you put it?

	1968	1969	1970	1971	1972
Home (raw)	13820·7	12287·4	11879·4	10126·6	9180·1
Imported	16736·6	17911·9	19478·0	16946·8	17065·3

Figure 19 (table, above) Iron ore consumption, 1968–72: consumption in blast furnaces, sinter plants and steelworks (thousand tonnes)

Figure 20 The iron ores of Britain

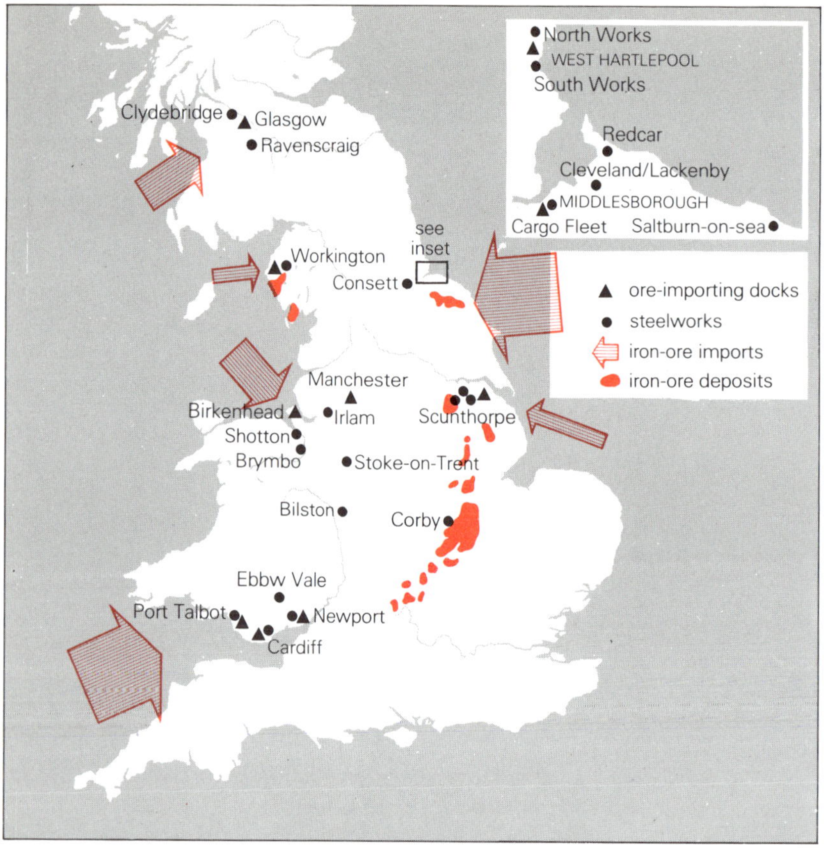

Let us take the same location problem a bit further. High-grade ores, which means ores containing a lot of iron, can be imported from abroad.

19 Where would you put a steelworks which was both near a coalfield and on the coast to import foreign ore? You have already seen a photograph of one (figure 18). Which other areas could you choose? You need only give a rough idea since you have not enough information to make a detailed choice.

20 Can you find any places which would be fairly near a coalfield and near *both* low-grade and imported ore?

Figure 21 (left) Integrated steelworks in Britain

Figure 22 (below) The interior of this modern steelworks at Scunthorpe gives an idea of the huge size of steelmaking equipment (find the man in the picture to compare sizes)

Figure 21 shows where the **integrated** works (those which make iron, steel and strip all in one works) are located. As you can see, some are:

(i) on or near coalfields and on the coast;
(ii) on the coast and near both coalfields and ironfields;
(iii) on the inland coalfields.

21 Why do you think there are still steelworks in the third type of location when it would be cheaper to transport coal to the orefields, or to the coast, than to carry iron to the coalfields?

Figure 23

Steelmaking problems

The answer to question 21 is not simple, but one of the main reasons is illustrated in figure 22.

(i) Imagine how much it would cost to knock down all the equipment shown in figure 22. It is cheaper to carry the ore to the steelworks, even though it costs more to carry than coal. This is called **industrial momentum**.
(ii) Many thousands of people work in a big steelworks. Rather than put all these people out of work, it is better to make less profit but keep up employment in the area. This is a **social** reason.

In fact the second reason is much more important. Many of the coalfield works are so old that it probably would be better to pull them down. The countries like USA and, particularly, Japan which started to make steel later than Britain, have the advantage of more modern equipment. To meet this overseas competition, the best answer would probably be to close down most steelworks and concentrate production in a few. These few would be huge, thus making economies of scale, and would all have deep-water harbours for imported ore. Ideally, they would be near the home orefields as well.

22 If you had to decide on *three* existing steelworks, in which *all* British steelmaking would take place, which three would you choose?
23 Explain the reasons for your choice. Try drawing a map to show the advantages of each location chosen.

Of course, because of the massive unemployment which it would cause, this cannot be done, but the British Steel Corporation feel that because of foreign competition there must be some concentration. Some of the oldest works are being closed and some of the larger coastal ones are being enlarged.

There will naturally be more jobs at the enlarged works, but not enough to make up the losses. In any case the jobs would not be in the same region as the unemployment. The steelworkers and their families who are threatened are naturally not very pleased. They think that the BSC should modernise existing factories in order to compete better with overseas rivals.

Before leaving steel, there is one more cause for concern.

At first sight figure 23 may appear to be a poor photograph.

24 Why do you think it is so hazy? Is it because it is a bad photograph, or might there be some other reason?
25 Think of three reasons why the tree has no leaves. Which of these is likely to be the real one? In other words what was the most likely purpose of the photographer who took this picture?

Steelmaking gives off a good deal of smoke and various gases. Moreover it uses a vast quantity of water, and when this water is returned to the rivers it is not very clean. The situation is not as bad as in the description of the Swansea Valley earlier, but it is not very nice, even now, to live near a steelworks, and there are not many fish in the rivers. What is to be done? Steel is so important in the nation's economy that pollution can never be prevented entirely. However, it is good to know that the BSC is spending several million pounds to reduce environmental damage.

Location of the shipbuilding industry

26 Before reading on, try to think

Figure 24 *The hull of a 'very large crude carrier' under construction in the Swan Hunter yards at Wallsend*

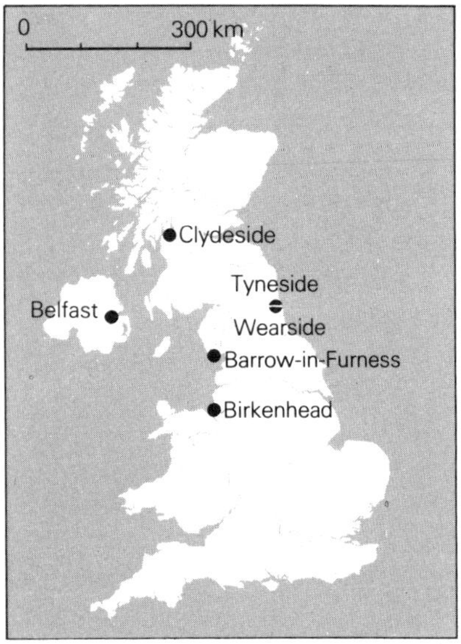

Figure 25 The main UK shipbuilding areas

what would be the main location factors for the shipbuilding industry. You need:

(a) A place where the hull can be built and then launched into deep water. In figure 24 you can see the partly-constructed hull of a big ship which will eventually be slid down into the river in the background.

(b) A place to complete the building in sheltered deep water. A ship is so big and heavy that the inside and superstructure have to be built after launching. The building area may be a sheltered part of a river, as for the ship in the background of figure 24, or it may be a specially constructed building dock.

(c) A supply of steel as close as possible to reduce transport costs.

(d) A supply of fuel, because the construction of a ship uses enormous amounts of energy.

27 With this information, and before looking at figure 25 where, roughly, in the UK do you think the main shipyards will be found?

Look closely at figure 25 and see if all the shipbuilding areas meet all four location requirements listed above. There is one in particular that does not. This is Belfast.

Your attention has been drawn to Belfast to show that location is not always as simple as it may so far seem. In the case of the Harland and Wolff shipyards, there are less obvious factors at work, such as having good management and brilliant ship designers at a time when shipbuilding in the UK was expanding rapidly.

Unfortunately, shipbuilding is no longer growing. Like steelmaking, the industry in Britain has suffered from being first.

Shipbuilding problems

Look at figure 26.

28 Which country launched the greatest tonnage of ships in 1938?
29 What percentage was this of the world total?
30 Which country was the leader in 1971?
31 What was that country's percentage of the world total?
32 What was the UK's percentage of the world total in 1971?

District or Country	1938	1956–7	1961	1971
Tyne and Blyth	157	265	208	194
Wear	169	211	260	265
Tees and Hartlepool	99	164	117	109
NORTH-EAST TOTAL	425	640	584	568
Clyde	444	398	332	284
UK TOTAL	1030	1383	1192	1239
Sweden	166	575	780	1645
Germany	481	1100	900	1617
Norway	55	205	390	798
Denmark	158	156	200	733
France	47	375	370	1216
Japan	442	2023	1650	11 184
WORLD TOTAL	3033	5995	7500	21 700

Figure 26 World shipbuilding (launchings in thousand tonnes)

The British shipyards are too small and too old-fashioned. Methods of working and labour relations are also out of date. Do you know what is meant by 'labour relations'?

Ideally, vast new shipyards should be built – economies of scale again. These should employ fewer people but make more ships. But, like steel, so many people are employed in shipbuilding that closing too many yards would cause too much hardship. The answer has been to merge many smaller yards, to enlarge others and to have help from the government. But these are not really satisfactory answers to the present-day problems of the UK shipbuilding industry.

Imagine you are a skilled shipyard worker faced with redundancy. You live in the Tyneside area. If your shipyard closes, what will you do? Some of the choices open to you are as follows:

(a) Stay on the dole and hope that a new shipyard would offer you a job later on.
(b) Find a new job in a different industry. If this is your choice, it presents several more difficulties:
(i) You only know one job well. Should you accept an unskilled job for less money or spend time training for a new one?
(ii) Should you move out of the area? This would mean leaving all your relatives and friends and the neighbourhood where you feel at home. It would also mean a better chance of getting a skilled job.
(iii) You will have worries about your family. Should you move house while still unemployed? How will you find somewhere to live? Do you take your wife and children, or do you leave them behind while you look for work? If you decide to settle in a new area, how do you look after your parents and in-laws, who are old-age pensioners and not too well?

We shall see more of this problem later. For the moment, you can see now why it is not so easy just to close down major works, even though in the long run the country as a whole would probably benefit.

REVISION

33 Make a list of the location factors that have been mentioned so far. When you have made your own list, turn the book upside down to check your answers.

1 Reduction of transport costs. (This includes nearness to raw materials and fuel and nearness to markets.)
2 Industrial momentum.
3 Industrial 'chain reactions'. (The attraction of industries of similar kinds.)
4 Economy of scale.
5 Social needs.
6 Government action.
7 Chance. (Like the luck of having a good manager, or a brilliant ship designer.)

The 'heavy' chemical industry

'Heavy' chemicals are those used in industry, rather than those you would find in a chemist's shop. The industry was originally located near bulky raw materials. The chief ones were salt, anhydrite, gypsum, limestone and coal. The coal was not used for burning, but as a source of hundreds of different chemicals.

Although the industry first started near raw materials, it developed very rapidly. Perhaps more than any other

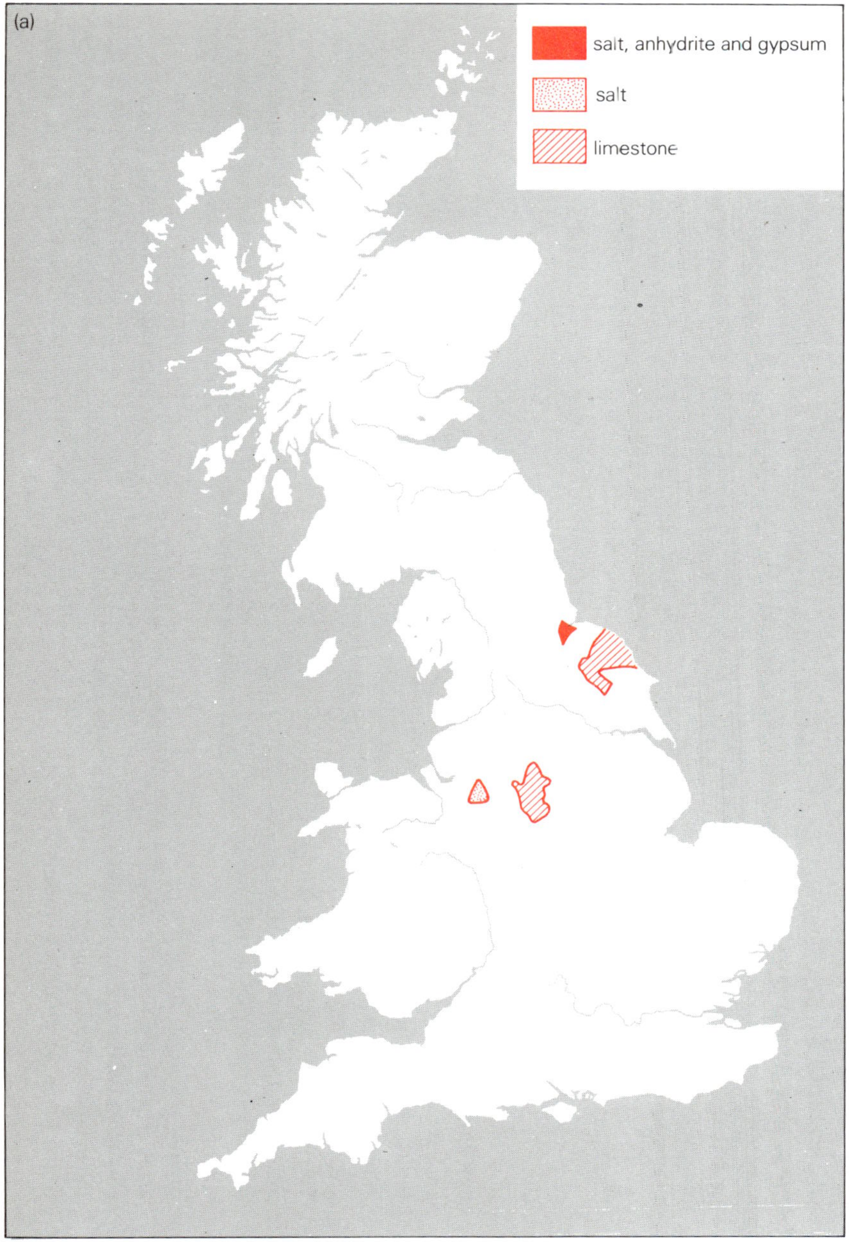
(a)
salt, anhydrite and gypsum
salt
limestone

(b)
Head Office
Division etc. Headquarters
Main factories and Research Stations
Grangemouth
Leith
Glasgow
Ardeer
Dumfries
Annan
Kilroot
North Tees
Billingham
Wilton
Heysham
Horton
Fleetwood
Harrogate
Thornton
Clitheroe
Darwen
Huddersfield
Shevington
Immingham
Widnes
Doncaster
Runcorn
Oldham
Ellesmere Port
Manchester
Penrhyndeudraeth
Hyde
Northwich
Derby
Buxton
Winsford
Macclesfield
Alderley
Coventry
Bromsgrove
Stowmarket
Gloucester
Stevenage
Pontypool
Welwyn
Slough
London
Severnside
Jealott's Hill
Yalding
Fernhurst
Kennall Vale

Figure 27 (page opposite) (a) Some chemical industry raw materials (b) The distribution of ICI factories gives some idea of the spread of the chemical industry

industry it causes chain reactions. This is because very often one chemical is the raw material for many others, and these in turn for others again. The products are enormously varied. Just a few are: pharmaceuticals (chemist's shop products), fertilisers, plastics, paints, explosives, textiles.

There is one other major location factor to consider in today's chemical industry. Oil and natural gas contain much the same chemicals as coal. They are also cheaper to process and transport. Thus many chemical industries are to be found near oil refineries and the great oil companies are now big chemical producers. In some cases, chemical companies have built their own oil refineries.

Even though there are now petro-chemical works near many oil refineries, and even though many smaller firms have merged, fewer, and even larger, factories (as in steel and shipbuilding) would lead to greater efficiency.

However, because of the development of pipelines, large factories are perhaps not quite so important in chemicals as in the other industries. Oil, natural gas and many chemicals can be pumped through pipelines. This method of transport is much cheaper and more efficient than any other way, so it matters less if the goods have to be carried some distance.

34 What do you think is meant by 'petrochemical' works?

At this stage we must consider another factor affecting location. Very large companies, with factories in many different areas, have to make location decisions with the whole company in mind, not just one factory (figure 27 (b)). This certainly makes location much more difficult to study (see next chapter).

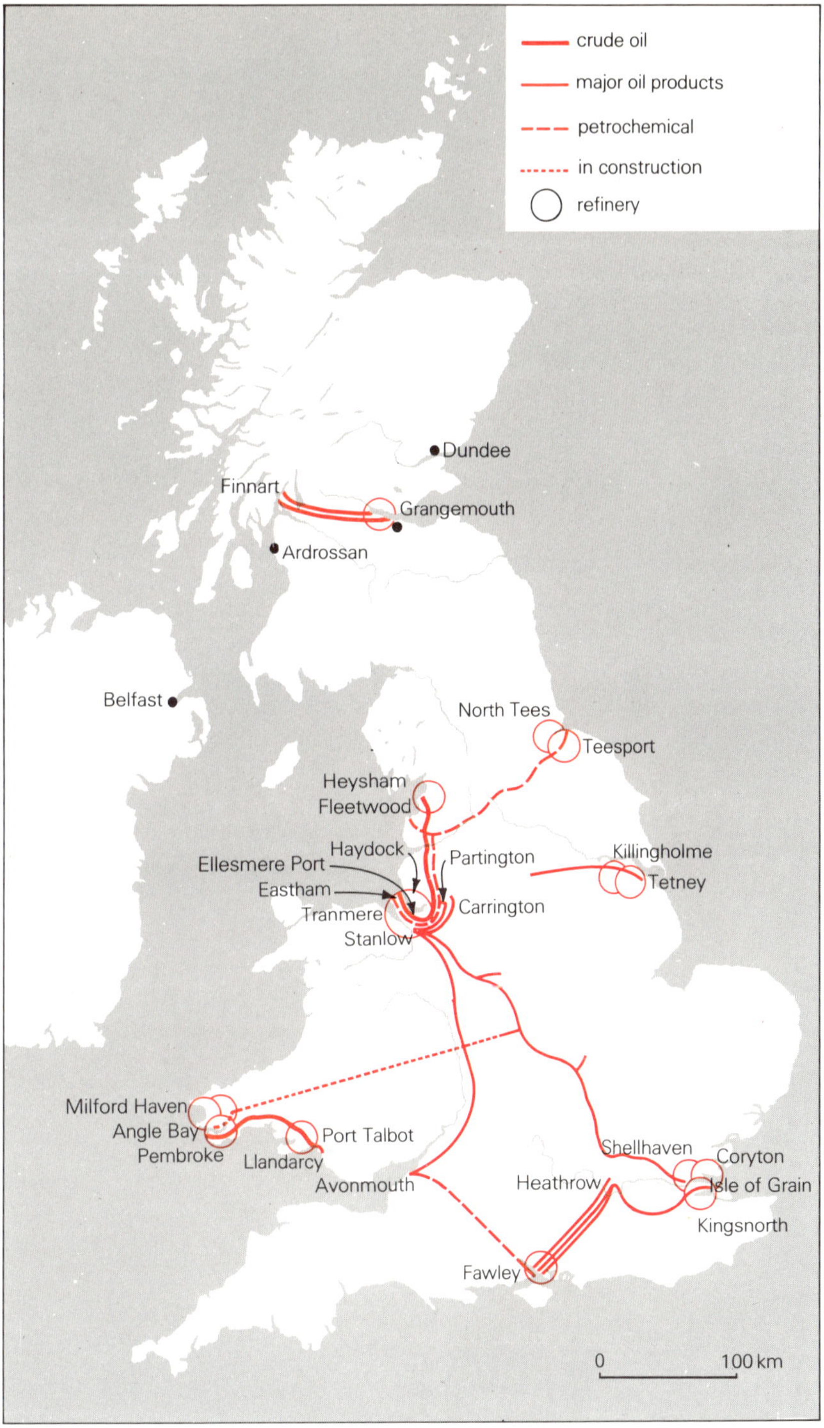

Figure 28 Oil refineries and pipelines

Pollution again

Figure 30 shows how very complicated a chemical works is. One writer has described it like this:

'From the impressive collection of fractioning columns, pressure vessels,

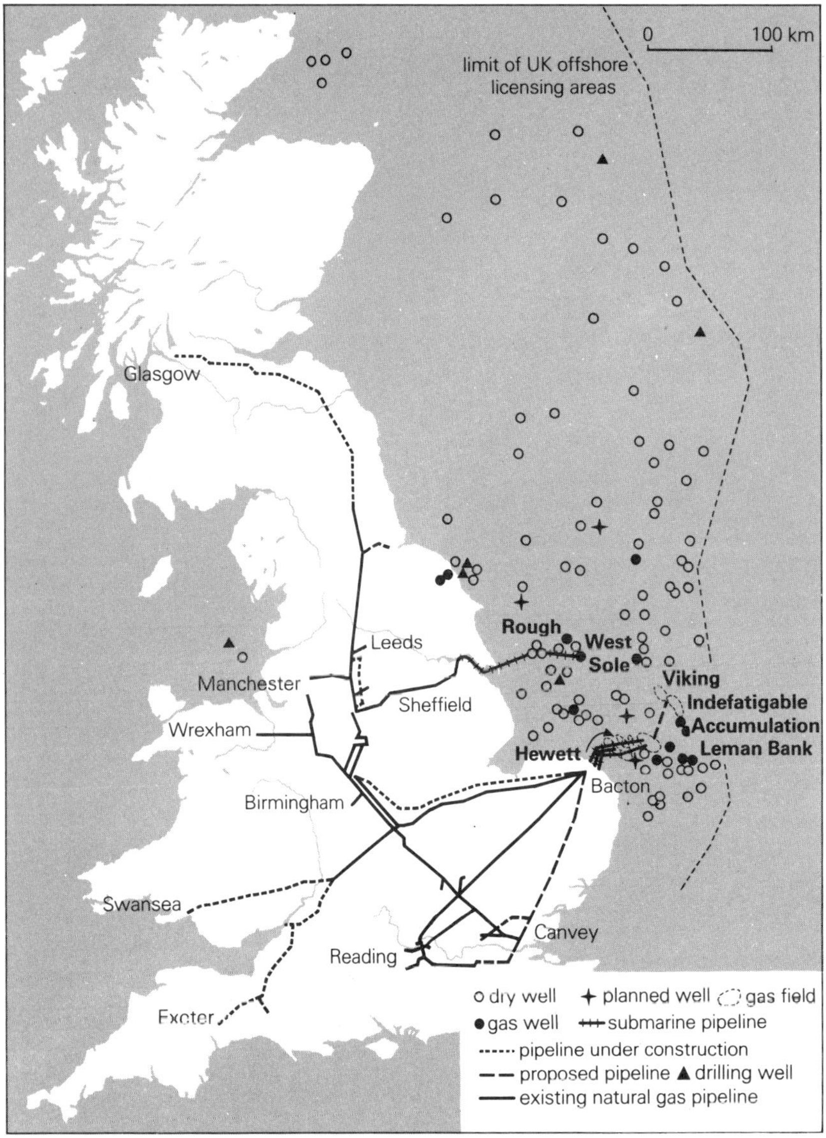

Figure 29 Major natural gas pipelines

domes, tanks, cooling towers, chimneys and the convoluted pipework that links them, steam, multi-coloured gases and a range of odours escapes at point after point. The movement of waste into the atmosphere is matched by that which finds its way by various sluices into streams and estuaries. Tall chimneys along the Manchester Ship Canal and the Mersey pour out their waste over the low ground. Haze, or thickly charged mists, drift high over industrial Teesside from Billingham and . . . Haveron Hill and Port Clarence huddle on its lee side as filthy, dust-strewn communities.'

35 Some of the features mentioned in the first sentence of this extract can be seen in figure 30. Can you find out which is which?

36 Which areas of Britain (not individual places) are badly affected by pollution from chemical works?

Unfortunately, it is not just the works themselves that pollute. The limestone quarry in figure 70 (page 57 in chapter 5) is right on the edge of a National Park. There are many other mineral workings in or near beauty spots, which may *look* ugly, even if they are not dirty.

As with steel, the chemical industry is too vital for us to be able to stop damage altogether. At the same time, it is important to realise that quality of life is as important as economic efficiency. More and more money must be spent on limiting harmful waste and on hiding unsightly equipment. The big chemical and oil companies are spending many millions on control. Even so, people need to be aware of their own interests. You can help, as you will see later.

Figure 30 Inside a chemical works

Distributive industries

In looking at industries which need to keep transport costs to a minimum we have so far considered only the three chief heavy industries. But it is not only bulky, low value goods, like iron ore, that are expensive to transport. Anything that will perish or decay needs either very fast (and thus expensive) transport, or else it needs preservation equipment, which also costs more. Goods which are packed in heavy, fragile containers (e.g. bottles, which are both heavy and fragile) are also expensive to transport. Thus we have a group of industries whose location is influenced by the cost of transport to the people who will buy them – their market.

37 Try to make a list of industries which would be found near their market and which would fit into the types described in this paragraph.

One example would be a milk bottling factory. Obviously factories that *preserve* the milk, by turning it into cheese or by canning it, will be less governed by transport costs to market. Their produce does not go bad so quickly as liquid milk.

PROJECTS

1 If your school has any pollution-testing kits, you could ask your teacher to help you organise a pollution survey. When organising a survey, you have to be careful not to let your opinions influence the result. You will remember from the last chapter that people 'see' the world differently. It would be quite easy to 'prove' that your area was polluted if you did your tests in an area which you knew to be contaminated. However, someone who wished to 'prove' the opposite could doubtless find a clean place to carry out the tests. For this reason, it is necessary to organise a survey so that you cannot make the result fit in with your ideas. This is called being *objective*.

A good way to organise an objective survey is by **point sample**. This means measuring the pollution at a number of different points. A simple way of choosing points is to use the intersections of grid lines on an OS map. If you only wish to cover a small area, further grid lines can be drawn between those already on the map.

In choosing the sample points in this way you are always likely to find that some points are inaccessible, in the middle of a field of crops or on the top of a gasholder, for example. In this case, you can use the nearest accessible point, so long as it is the *nearest*, not one that you just happen to like.

By doing a point sample survey, you cover an area which should be big enough to find out whether there is any difference between one part and another. If there are differences, you can begin to look for reasons to explain them.

Ideally, such a survey should be carried out so that all the points are surveyed at the same time. However, this may not be possible because there may not be enough people or enough testing kits. Although pollution does vary from one minute, or day, or week, to the next, it is possible to do a worthwhile survey by doing the tests at different times. This is because various **indicators** of long-term pollution do not change. Such indicators include the plants and animals in the water, on the ground, or on buildings.

Carrying out a survey such as this is a fairly big job, involving perhaps a whole class and a lot of organisation. However, if as a result you could produce a pollution map of your area, you could well be doing a most useful service to your community.

2 (a) Using the statistics in figure 26, draw line graphs to show the changes in ship tonnages launched for each of the different countries. Use a different-coloured line for each country.

(b) Using a reference book, such as *Geographical Digest* look up the six leading crude steel producing countries. Are they the same as the six leading shipbuilding countries?

Draw a column graph (histogram) to show the steel production for each country.

(c) You may wonder why exercise (b), above, did not ask you to draw a line graph or why it did not mention chemicals. This is because statistics over a period of time are only found in the bigger reference books, such as the *United Nations Statistical Yearbook*. You may have this in your school library, or it may be in your local reference library. It is much bigger than the *Geographical Digest* and the number of pages of figures, the title in both French and English and its size may put you off. However, if you are prepared to have a go, it only needs common sense. Look up the main heading 'Manufacturing' and the subheading 'Crude Steel' in the table of contents. The contents are in English on the left side of the page, French on the right. Then look up the table and find the six largest producers at the most recent date given. Simple!

Chemicals are not so simple. There is no single figure given for them, but lots of different products. However, the production of sulphuric acid is often taken to be a good reflection of the total chemical production of any country. You can look up sulphuric acid in the *UN Statistical Yearbook* in the same way as crude steel. Why not have a go? Do the same exercise as in (a) and (b).

3 No . . . let's make it there instead

1 Look carefully at the pictures in figure 31. Can you think of any connection between the different events shown?

Figure 31 (above and opposite)

In October 1973 a new war between Israel and her Arab neighbours flared up. The war ended with an uneasy ceasefire, during which the Israeli army occupied parts of Arab countries which it had captured in this and other wars.

You may wonder what this has to do with the geography of the UK. We shall see here, and again elsewhere in the book, that Britain's economy is dependent on what is happening in other parts of the world. Some Arab countries are among the world's chief exporters of oil. Although these countries were not themselves in the war, they wished to help their allies who were. So they cut down the amount of oil they exported to Europe and to other countries. The idea was that these countries would then try to persuade Israel to leave the Arab territories.

For the UK the result was a major oil shortage. Obviously this meant a shortage of petrol for cars, but what other effects would there be? To answer this question, it is necessary to look at what was happening in Britain.

Changes in power supply

Look at this table:

Figure 32 Employment in coal mining and number of coal mines

	1965	1970	1975
Number employed in coal mining	450 400	285 100	246 700
Number of coal mines	504	293	241

2 What has happened to the number of people employed in coalmining?

3 By what percentage has the number been reduced?

4 What has happened to the number of coal mines?

5 By what percentage was the number reduced between 1965 and 1975?

Figure 33 shows miners at the coalface of a fairly modern mine. Have you ever been down a coal mine? If not, you should try to

Figure 33

Figure 34 UK inland energy consumption, percentage share of total

	1949	1959	1969	1975
Coal	90·4	76·5	50·7	36·4
Oil	9·2	22·6	42·7	42·6
Natural gas and colliery methane	—	0·1	2·7	17·3
Nuclear electricity	—	0·2	3·3	3·3
Hydro-electricity	0·4	0·6	0·6	0·4

arrange a visit if you possibly can. No description can really bring to life the dust, the noise and the frequent dampness in even the most modern mines. In the older mines, even with the use of machinery, conditions are much worse.

What, then, is so terrible about closing down an old mine and concentrating on newer ones where the work is a good deal easier?

The chief difficulty is that coal-mining, perhaps more than any other industry, gives employment to a whole community. So if you close one pit, you may throw hundreds out of work, because there is no other industry in the area to change to. The problem is very similar to those of steel and shipbuilding explained in chapter 2, but worse.

Despite the social hardships many mines have been closed, as you have seen, and many fewer men are employed in coalmining. Why has this rundown been necessary? Look at the table in figure 34. Look first at coal. See how the percentage of UK power produced by this fuel has declined.

6 Which two fuels increased their percentage most in the time shown?

In 1970, oil overtook coal as the chief source of energy in the UK. Now perhaps you can see how the shortage of oil in the winter of 1973/4 affected the economy. Nearly half of all the oil is used in industry and almost a fifth is used to generate electricity, much of which is used in industry. In that same winter there was a miners' industrial dispute, so that even the reduced amount of coal we have mentioned was not being produced.

The shortages of oil and coal together, meant shortages of power for industry, which had to go on to

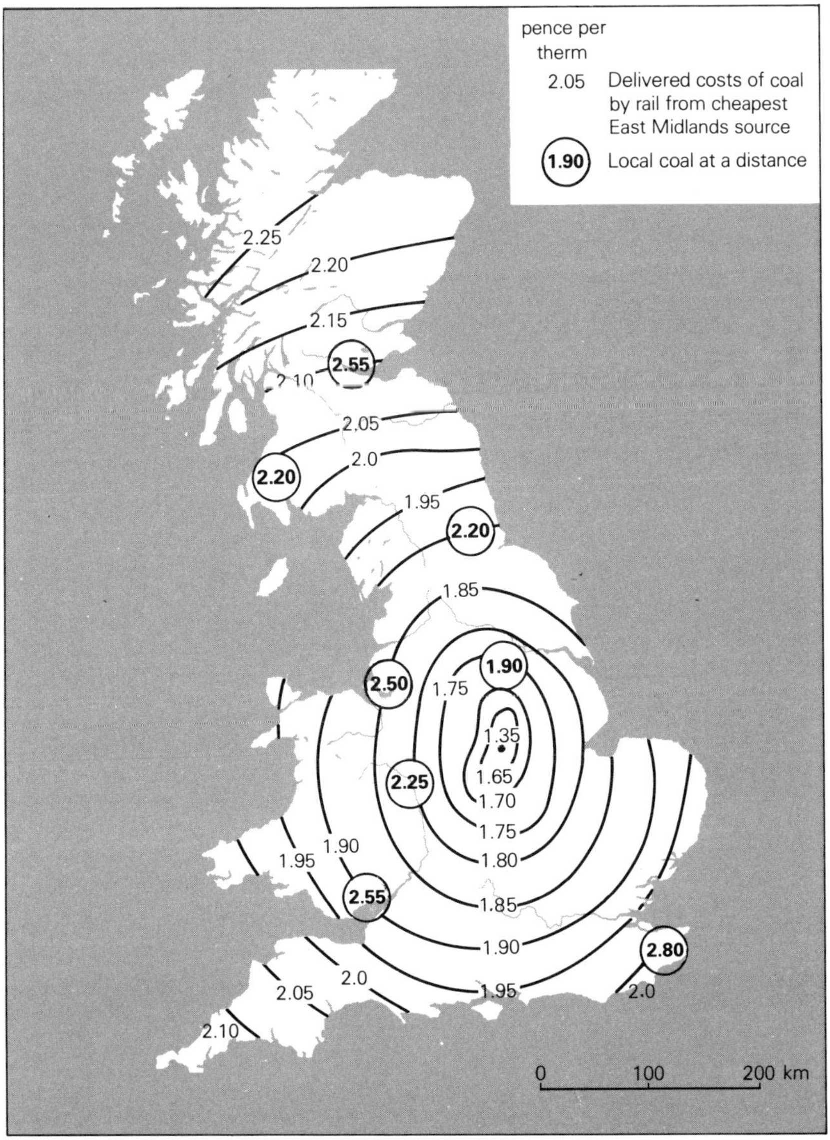

Figure 35 Map to show costs of coal in different parts of the UK

working a three-day week. They also caused electricity blackouts in offices, shops and homes, not to mention long queues for petrol.

The changing location of coal-mining

How does all this affect location? Why should the change from coal to oil affect the *place* where energy is produced? One obvious reason is competition.

If oil is cheaper than coal, industries which use power will wish to change to the cheaper fuel. One way to cut the cost of coal is to close old, inefficient mines and to increase output at new mines.

At first sight, figure 35 may seem a little difficult, but it is not really. The circles represent coalfields. Look at an atlas to see which coalfield is which. The numbers in the circles represent the cost of producing coal on that coalfield. In fact, to make it easier to compare one area with another the costs are given in pence per therm. A therm is a measure of the heat which the coal could produce.

7 Which coalfield produces the cheapest coal?

8 In which two coalfields is it most expensive?

The lines on the map show the cost of the coal added to the cost of the transport.

Look at the line which has 1·85 on it. This means that anywhere on that line the coast of coal and transport amounts to 1·85 pence per therm. You will see that the line passes through the Lancashire coalfield where the cost of *its own* coal is 2·50 pence per therm.

9 Is there any place shown on the map where it is cheaper to use coal from the nearest coalfield, rather than to buy coal from the East Midlands?

10 Name the coalfields which produce more expensive coal than that from the East Midlands, added to the cost of transporting it. Which two coalfields are most likely to have suffered less unemployment in coal mining?

The need to produce coal more cheaply has led to most coal being produced in the East Midlands area. There has been a *change in location* of the chief coal-producing areas. Even within the more expensive coalfields there has been a change in location, from older to new mines. Another way of reducing costs which will also produce a change in location is to open a few completely new mines in places where it is easy to get at the coal. One such mine is planned near Selby, to the east of the present Yorkshire, Nottingham and Derby coalfield.

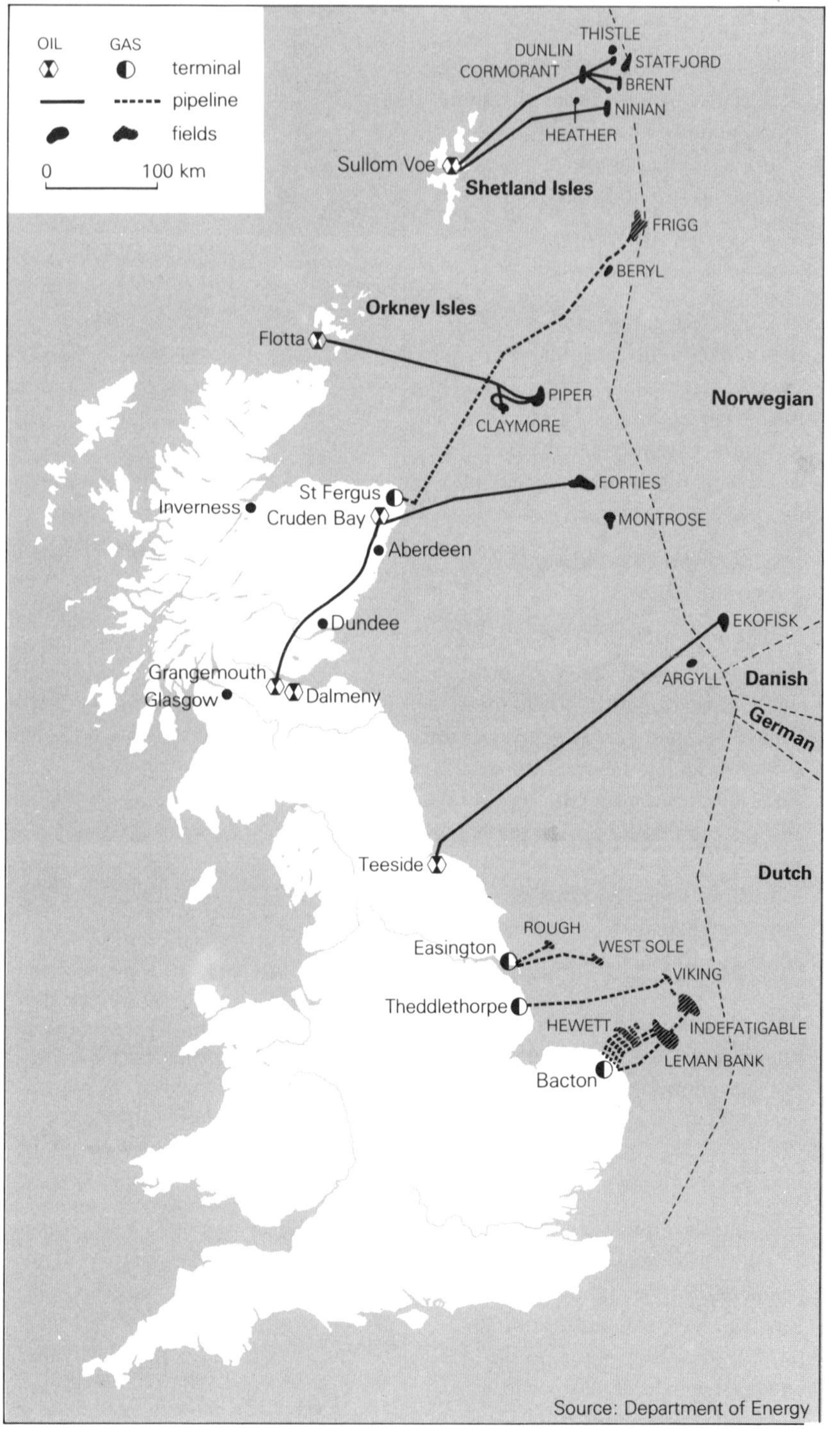

Figure 36 Main oil and gas fields in the North Sea

North Sea gas and oil

The Arab oil cutback also made the importance of some other changes more obvious.

Look again at figure 34 and at your answer to question 6.

11 Which two forms of power appeared for the first time in 1959?

North Sea gas is now piped into many households, and major industries have signed contracts for long-term supplies. The fact that gas now comes from under the North Sea means that another change in location has happened.

12 Do you know where our gas came from before?

This change is partly due to luck; the chance that the gas was actually there. It is also partly due to improved technology. Man now has the ability to find and extract the gas from beneath the sea. But gas is not all that was found beneath the North Sea.

Imagine the importance of finding huge supplies of oil on our own doorstep! Exploration by the big oil companies has revealed that under the northern part of the North Sea there is a geological basin containing perhaps enough oil for all Britain's needs. There may even be a surplus.

Prosperity for Scotland?

Building the equipment to produce North Sea oil helps to solve another problem. For years the Highlands of Scotland have suffered from high rates of unemployment and the young people have moved out, looking for prosperity elsewhere. Now the oil industry, and construction for it, is providing thousands of jobs. People are now speaking of 'Scotland's Texas' and of Aberdeen as the 'oil capital of Europe'.

Read carefully the extract in figure 37.

Scotland's Texas is taking shape

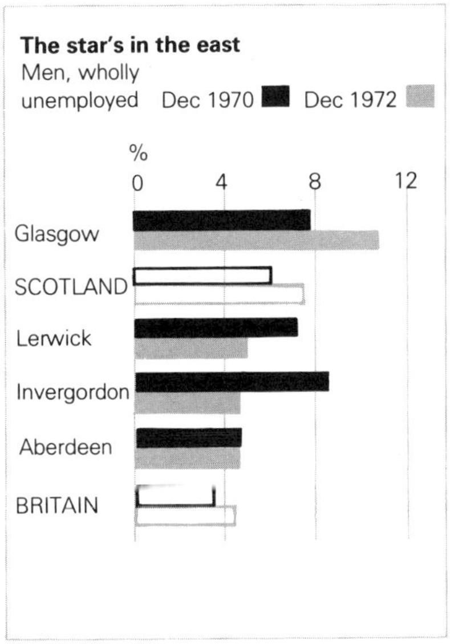

Scotland's Texas is taking shape. There are 5,000 Scots now who have the North Sea oil industry to thank for their jobs, and this is just the start. If no more oil were to be found in the North Sea, the industry at its present size would still be employing 12,000 people by 1976. But if oil continues to be found at the present rate it will become one of the biggest employers in Scotland by the end of the decade.

So far, the impact has been confined to the north-east corner of Scotland, with Aberdeen as its capital, and parts of the Highlands and Islands. The effect on unemployment has been such that there are now worries about labour shortages. In the Cromarty Firth area local businesses, especially builders, have more orders than they have ever known, but every day they lose men to the oil industry. In the Shetlands a bakery has actually had to close because it could find no one to do the work.

The greatest effect to date has been psychological. For generations, northern Scotland had tended to be written off as a depressed peripheral area, good only for crofters and tourists. Every year, hundreds headed south. Now all that has changed. Relative backwaters like Invergordon, Lerwick and Peterhead have suddenly become the hub of a great international industry.

Emigration has dried up, and immigration is beginning. Many of the first immigrants are returning Highlanders. The Highlands and Islands Development Board in Inverness has the names of 10,000 people who left the area but now want to return; the board has its own 'counterdrift' plan to bring them back. It is not just that the area needs more people, but that the migrants tend to be the young, the able and the ambitious. They are the stuff of economic growth.

Growth, jobs and industry are not to everyone's taste. There is a strong conservationist lobby that fears oil will destroy Scotland. Its members foresee a landscape scarred by industrial sites and beaches polluted by oil spills and dead sea birds. But above all they fear that the oil will not last. It will run out in 40 years, they say, and the industry will disappear, leaving behind derelict plant and mass unemployment.

These chiefly middle-class fears are mainly expressed by people who already have congenial jobs and a good income. They should not be ignored just because of that. Special squads will be needed to deal with spills; care needs to be taken over the siting of industry. But any idea that the oil will rapidly run out is an illusion. The search for, and the discovery of, oil in the North Sea will go on for years. Even if it is exhausted in 50 years' time, what other industry has so long a guaranteed lifespan?

Figure 37

Start on the " largest hole on earth " at Nigg Bay, Cromarty. The rig dock has now sucked in all local tradesmen

What price a way of life?

Marshall Pugh goes home to Scotland to see the effects of the oilmen's dreams

WILD GEESE are fleeing faster than the tourists as winter comes to Wester Ross on the rain from the sea. Red stags are roaring in the rut instead of charabancs on heat while robins hug the hedges ···red caravans.
···v, local residents ···l live their own ··· will be no peace ···ss this side of
···le of Lochalsh to···t a public meeting ··· the steel and con···tion camps planned ···nd Loch Broom

They were relieved to hear Lord Cameron, chairman of the advisory body for the Highlands and Islands Development Board, remind the Government that it wasn't possible to determine what, if any, overall policy was being followed.

They shared the " secret " that Ross and Cromarty Council will refuse to provide essential roads, schools, fire brigade, h··· and ···icemen if oil c··· ···ds are " railroade··· ···nst their wishes ··· the men of ···

named. As one of them said: " What's the use? We're spitting into the wind and that wind is dead against us."

Need this be true? If you can suffer to translate the polysyllabic illiteracy of the official predictions and announcements, the issues become clearer.

For a start the oil industry as a whole ··· give a ligh··· fill wheth··· ··· towers built i··· ··· or

long ago, and drained the county of tradesmen so that people in Cromarty must now send 20 miles for a plumber. Nigg is already recruiting as far south as Newcastle.

Nigg was once known as the largest hole on earth. It is now the world's largest graving doc··· (for building steel platforms and a standing answer to ··· claim that Wester R··· restored after the ··· towed to sea. Th··· at the dock the ···ven in silence, ···ested the ··· The wi··· ···are

13 Look up the following places in an atlas:
Aberdeen; Cromarty Firth; Invergordon; Lerwick; Peterhead.

14 'The Highlands and Islands' are also mentioned in the extract. You may not find them under this name in the atlas. See if you can find out where they are.

15 Look at the graph concerning unemployment in the extract:

(a) Three towns mentioned in the article are shown in the graph. Was there as much unemployment, or less, or more, in 1972 compared with 1970 in each of the towns?

(b) Was there as much unemployment in the towns in 1972 as there was in Scotland as a whole?

(c) Now compare the 1972 unemployment with the figure for Britain as a whole.

(d) Why do you think Glasgow has been included on the graph?

16 The word 'counterdrift' is used in the article. Explain in your own words what you think it means.

So far we have looked at the advantages of North Sea oil to the Highlands and Islands of Scotland, but as usual, solving one problem creates another.

Figure 38

17 What are the three problems named in the next-to-last paragraph of the extract?

Some people see the problems as even greater – look at figure 38.

18 What do you think is suggested by the headline in figure 38?

Massive rig-building docks, pipeline terminals, possible oil spillages, together with the arrival of thousands of workers, are quite capable of ruining beautiful scenery, and Scotland has some of the finest unspoiled landscape in the UK. On the one hand the possibility of making Britain self-sufficient in oil and providing work for thousands cannot be neglected. On the other hand the wrecking of some of the finest coastlands cannot be ignored.

Oil refining

Most people know that the oil that comes out of the ground contains an enormous number of useful things. The process of getting these things out of the oil is extremely complicated and is called refining.

The main location factor at present for oil refineries is a deep-water harbour where the big tankers can dock safely. It might be expected that with less oil to come from overseas in future, the oil refining industry would move, perhaps into eastern Scotland.

However, oil refineries are very expensive to build. They are also relatively new. So their industrial momentum is very great. (If you cannot remember what is meant by industrial momentum, look back to page 23.) If North Sea oil produces any changes at all in the location of refining, it is likely to be simply that the refineries on the east coast expand more than those on the west.

The refined products can be carried cheaply by smaller coastal tankers, by road and rail tankers, or very cheaply by pipeline. It is mainly cheap transportation which will concern us in the rest of this chapter.

Energy from electricity

Coal, oil and gas are fuels which can be burnt to produce energy. This energy may be used to drive cars, cook your dinner, or work industrial machinery. However, although oil and gas can be transported fairly cheaply, coal cannot. Moreover, burning the fuel in the factory to produce your own energy is very inconvenient. Is there any way in which energy can be provided in a convenient form and perhaps allow cheap transport as well?

Look closely at figure 39. To the right of the big towers is a coal stock. This is burnt in boilers in the main building to produce steam. The steam is used to drive turbines. These are connected to generators which produce electricity. The steam is then cooled down in the eight huge cooling towers. In the background you can see a big river. Vast quantities of water from the river are used in the cooling towers. In the foreground of the picture you can see the pylons carrying the wires to distribute the electricity.

19 Where in England do you think this power station is likely to be? Remember, you need coal. Look back to figure 35 to see where the cheapest coal is. You need large quantities of water. Look in an atlas to see where there is a large river near the cheap coal.

When you have decided where you think the power stations will be, look ahead to figure 42 to see where they are.

20 What kind of fuel is likely to be used in power stations which are not near a coalfield?

When needed, power can be transported right across the country if it is boosted occasionally. There is therefore a network of transmission lines connecting the power stations with each other. If there is a shortage in one area, electricity can be brought in from another.

There is one other way of producing electricity that we must look at. Look again at figure 34.

Figure 39 Coal-burning power station

Figure 40 The electricity grid

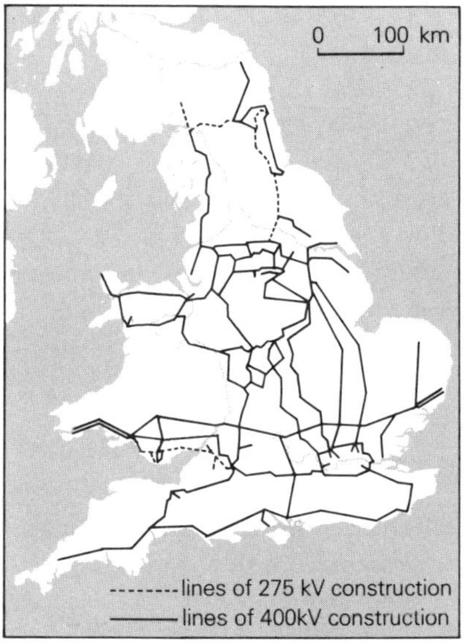

21 Which two sources of power have not yet been mentioned? Of the two, hydroelectricity contributes such a small quantity that we will not consider it any further at this point.

22 Do you think the power station in figure 41 is beautiful, or ugly, or neither in particular?

In this particular case the appearance is rather important. This building is in the middle of the Snowdonia National Park! A nuclear power station works in much the same way as any other. The difference is that very small quantities of fuel are used. Because of this, the world's supply of uranium should last much longer than the supply of coal, oil or gas. It also means that once the designing and building of the station has been paid for, the cost of producing electricity would be very low. Even though unexpected snags have arisen which have put up the cost, we can expect the use of this source of power to increase.

The chief problem with nuclear power is that if there were an accident, the effects could be disastrous. The likelihood of an accident is very small, but even so, account has to be taken of it. Therefore, most nuclear power stations are in remote areas. Huge quantities of water are needed (more than most rivers can provide) so the usual location is on the coast away from populated areas.

23 Although figure 42 is not particularly clear, how many nuclear power stations are obviously inland?

24 How many nuclear power stations are there altogether?

Coal mines, oil refineries and power stations all occupy large areas of land. All are big and unsightly. In the industrial areas we have got used to coalmines and even to open-cast mining, which is worse. Even conventional power stations, enormous and unsightly as they are, have perhaps been accepted by most people as part of the landscape. But now we have a power station in the Snowdonia National Park; consider the difficulties facing the people who made the decision to allow such a development.

Apart from producing energy, transporting it can also have its problems. The transmission lines in figure 43 which allow electricity to be moved so cheaply and conveniently

Figure 41 (above) Trawsfynydd nuclear power station

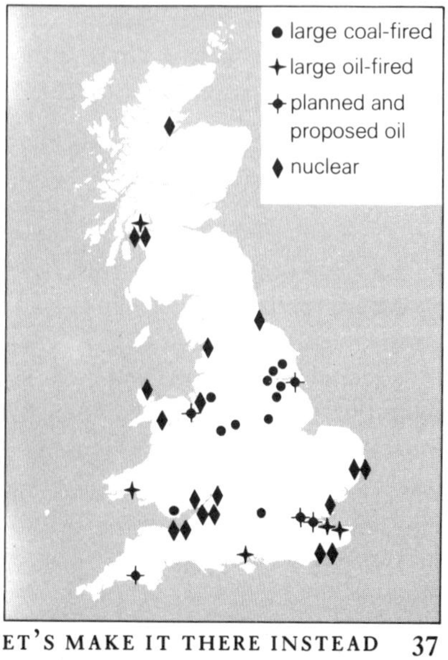

Figure 42 (right) Large power stations of Great Britain

are unsightly. Unfortunately, the cost of putting the lines underground is too great to make it worthwhile, except in the most important areas of natural beauty. Even here we are faced with the problems of opinions. Who is to say what is 'worthwhile' or what is an 'important' area of natural beauty?

Footloose industry

In the previous chapter we looked at industries which used large quantities of fuel. But not all industries are like this. Do you, or your family, have a transistor radio? Or a television set? Or a vacuum cleaner? Do you go on holiday every year? Maybe you go abroad by plane. It is likely that you have, or do, at least one of these things.

The industries which supply these goods and services, or which make the parts which are later assembled into them, are the fastest growing. This is because these are the things most in **demand**, not just in the UK, but in all the industrial countries.

25 Make a list of ten more manufactured goods for which you think there is great demand in the UK at the present time.

These industries do not need coal or crude oil. A great many need only electricity as a source of energy. They can therefore be located almost anywhere as far as power is concerned.

26 Can you think of any factors which might be important for their location?

Like any other industry, they have to have a market for their products. However, if the goods are fairly small in size and big in value compared with their size, like, say razor blades, or transistor radios, then it does not really matter how far they have to be carried. So long as the factory is near a convenient form of transport, the *distance* the goods have to travel does not make much difference to the price we pay for them in the shops.

Very often, the same industries use raw materials of light weight and high value, so transport costs are not very high here either.

So, what *does* influence the location? The very large group of industries where transport costs are unimportant are often known as 'footloose' because they can go where they choose. It is interesting to notice that even the footloose industries tend to collect in the same part of Britain, even though the exact location is not important.

Figure 43

The central belt of the UK

Most of the people in the UK live in the area shaded pink in figure 44. There are thus more people in this area to buy things. In other words, this is the largest market area in the UK. But it is also the area where the most **labour** is available to work in the factories and offices. In addition, it is where most people live who have special skills. Examples of these are scientists, electronic engineers, managers and **entrepreneurs**. An entrepreneur is a person who starts his own business.

Entrepreneurs in modern growth industries may be managers who think they can do better on their own; they may be scientists or engineers wishing to try out new ideas; or perhaps they are refugees from political oppression abroad who bring with them some savings and some 'know-how'.

The same shaded area on figure 44 is also the one with the densest road and rail networks. A dense network means that within this area, almost everywhere is near a road or railway line. Even the most modern forms of transport (motorways and freight-liner routes and terminals) are concentrated here. Look at figures 179 (p. 123), 186 (p. 126) and 194 (p. 130) in chapter 10 to see this.

We can see that the growth industries tend to like the central area. The presence of firms making similar things, of transport and of services like drains and electricity means that it is becoming more and more popular. Once the process had started, the forces pushing industry into this part of Britain became stronger and stronger. Unless something is done about them there will be little work for anyone in any other part of the country.

In fact something is being done, and has been for some years. We will look more closely at this in the next chapter. Meanwhile,

27 Can you think of any solutions to the problem?

28 Compare figure 44 with an atlas map of England. Which main cities are to be found within the central belt?

29 Two of these cities are much bigger than any others and are the chief centres within the belt. Which are these two cities?

As an example of industries in the central area, look at the map in figure 45 showing where the Lucas group of companies are. Most people have heard of Lucas, who make electrical equipment for cars and bicycles. But they, and their associated companies, make many other products as well – far too many to list here.

30 Compare figure 45 with figure 44. How many of the Lucas factories are in the 'central belt'?

Although Lucas make the kind of things characteristic of growth industries, they are probably not a very good example. This is because they are a very big company, whereas many of the really typical firms are small.

31 Why should a very big company not be a good example?

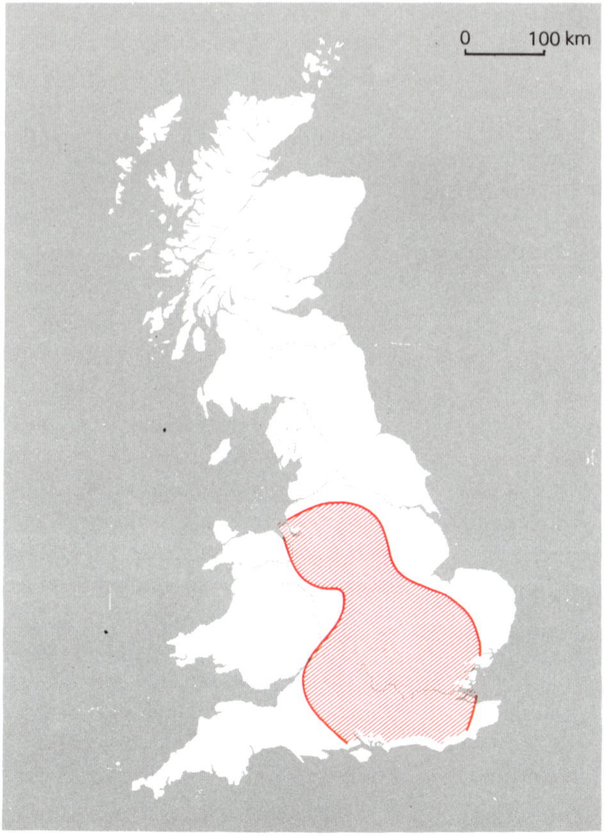

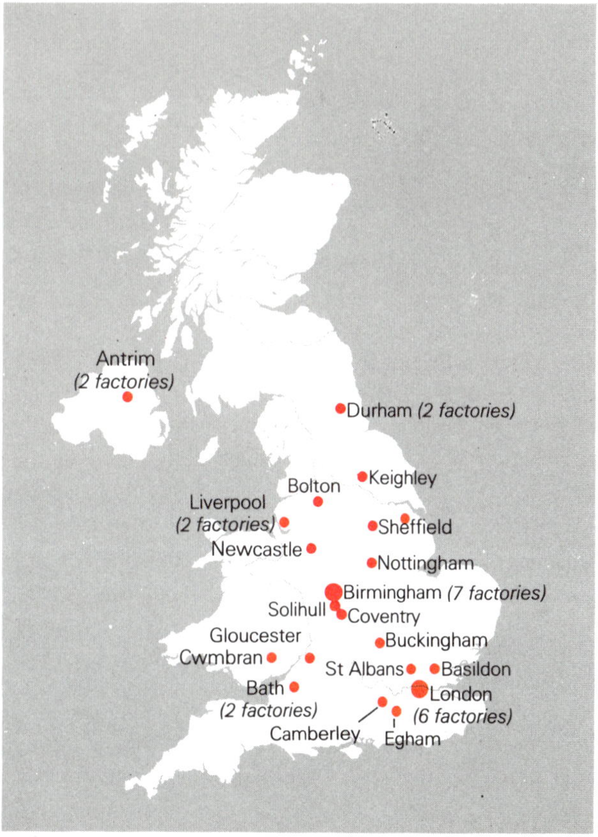

Figure 44 (below left)
Figure 45 (below right)
Lucas industries in the UK

Figure 46. The assembly line in a Ford factory

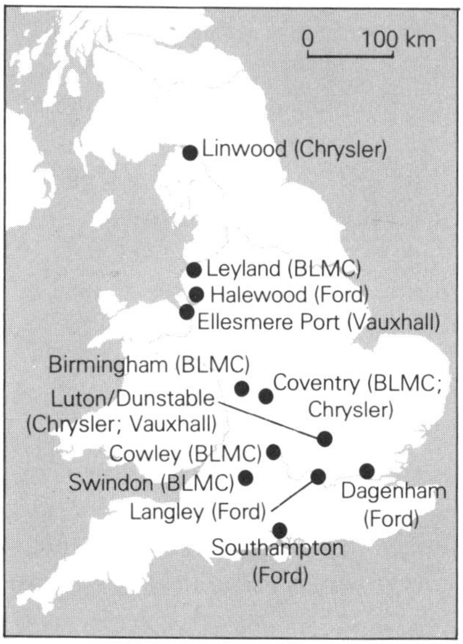

Figure 47 Motor vehicle manufacturing in the UK

In looking at the location of industry, we have been concerned with where to put a single factory. But nowadays more and more factories belong to the great corporations. In deciding where to put a single one of their works they may not need to consider that factory alone. More important would be the decision about how the factory fitted into the plan for the whole corporation.

This plan, or strategy, may apply to more than one country. Many of the big companies are international. In deciding where to put a new factory they may well be considering not 'where in Britain shall we put it?' but 'where in the world'! At this sort of level, Britain is so small that anywhere in the country is almost as good as anywhere else.

The motor vehicle industry

Look carefully at figure 46. The man in the picture carries out a few small tasks on each car as it moves up to him on the production line. He does the same thing to each car, hour after hour, day after day. Other men do other tasks, until eventually the car is finished. The production line does not stop; you cannot take a rest for a few minutes if you feel like it. But these workers are among the highest paid in industry. They are organised into unions to achieve high incomes and to try to make sure that the work does not become intolerable. The unions are able to negotiate with the management about pay and conditions.

32 Is this the sort of work you would like to do? When you have decided, take a vote among the class to see how many would like this type of work, how many would not. Give each other reasons for your decisions.

The motor vehicle industry is one of the most important in the country. The four chief companies employ some 316000 people, but in addition provide work for many hundreds of thousands in the industries which supply the car firms with vast quantities of steel, plastics, rubber and thousands of other products.

33 How many of the car manufacturing centres in figure 47 are in the central belt?

Since motor vehicle manufacturing is carried out by very big firms, some of them international, the factors affecting the location of the works are very complicated. Even so, two of the factors are worth a closer look.

We have seen that workers join unions, and this may be an important factor for the big companies where other things are equal. The management may try to avoid setting up a factory in an area (or country!) with a history of labour disputes. Instead it may try to find one where there is a tradition of good industrial relations.

Labour relations can thus be a location factor. Another is . . .

Government action

Many factories are located where they are because of government policy. When the companies decided to expand in Britain, they found that they were not permitted to build new factories near their existing ones, nor to expand them very much. This was because most of the factories were in the rich central belt (figure 44).

The government was anxious to encourage new industries outside the central belt, where unemployment was high because the traditional industries were employing fewer and fewer workers.

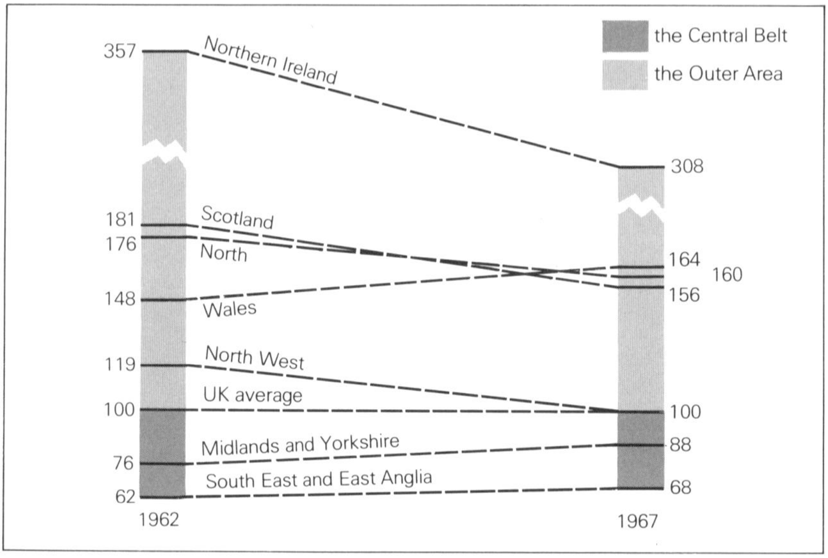

Figure 48 Average unemployment rates, 1962 and 1967

In figure 48 the average unemployment for the different regions is shown as percentages of the rate for the UK as a whole. There are two graphs in the diagram, one to show 1962, the other to show 1967. The dotted lines connecting them are to help to show the changes between the two years.

34 Was unemployment as high in 1967 as it was in 1962 in:
(a) Scotland?
(b) East Anglia?
(c) the South East?
35 Are the rates of unemployment in the central belt and in the outer area getting further apart, or closer together?
36 If you think they are getting closer together, from what you have just read, what might be the reason for it?

In persuading firms to set up factories in the outer areas, the government had two main weapons. First, no factories over a certain size could be built without an Industrial Development Certificate, and these were not usually granted in the 'rich' area.

The second weapon took on different shapes. Depending on which part of the 'poor' area the new factory was to be set up in, the **incentives** could range from tax relief to the provision of rent-free factories. They could also include loans and grants to buy equipment, or, if necessary, to build new factories.

In the case of the motor industry, most of the new factories have been sited near Merseyside, because this part of the 'poor' area was nearest to the component supplies in the Midlands. It was also near a port for easy export of the cars to overseas markets. However, one car factory is in Scotland. This is rather a long way from the traditional car area.

37 Why do you think the government would be pleased that the factory was built in Scotland?

It is not just national government which tries to persuade industries to move. Many local governments in the outer areas will give financial and other help to encourage industry to move in. Often they have a special department, offering publicity, information and other help.

In these last two chapters we have looked closely at the 'forces' which try to pull, or push, industries into certain locations. Most of these forces have been what might be called 'natural'. This is to say, they are economic forces which try to find the most profitable place for the industry. In the last century, these were probably the strongest forces. Even in this century, they are still very strong. However, at the end of this chapter we have looked at government **intervention**. Government action is necessary for social as well as economic reasons. In some cases it may force industry to accept locations which they would not otherwise do. Government action is now the most important location factor, and as time goes on it will probably become even more important.

PROJECTS

1 It may be that your school has some geographical games kits. Two of those published by Longman, *The Power Game* by Geoffrey Dinkele and *North Sea Exploration* by Rex Walford, would be useful to play at this stage if your school happens to have them.

2 Follow the share prices in the daily newspapers.

Exploration for oil in the North Sea is conducted by private companies, with government help. To raise money for their operations, companies sell 'shares'. Shareowners are entitled to a share of the profits of the company. Apart from shares being sold by the company, owners of shares often sell them to anyone who wishes to buy them. If the company is doing well, the profit from the shares is expected to be high and so the price of shares increases. The opposite happens if the company is not doing well.

Thus the change in the price of shares shows the prosperity of a company. Sometimes, because of national, or world, conditions, the prices of *all* shares rise or fall. In this case a prosperous company would be one whose prices did not fall as fast as the average, in a general decline. Similarly, prices rising faster than the average at a time of general increase would also indicate a successful company.

The share prices of the main companies are shown every day in the financial pages of *The Times*, the *Guardian* and the *Daily Telegraph*. Try to arrange for your class to take one of these papers for a period of time. Every day, note down the price of the shares of the firms involved in North Sea exploration and development. These figures can be drawn up on a line graph with the lines growing one point longer every day. Use a different coloured line for each company. The obvious companies to follow are the oil giants, such as Shell and BP, but their share prices are so often affected by events in the Middle East and other parts of the world, that they are perhaps not a good guide to the state of North Sea oil. Some less obvious firms, whose involvement in North Sea exploration makes their share prices a good indicator are: AHA; Assoc. News; Cawoods; Nat. Carbonising; Thomson Organisation; Tricentrol; Ultramar. The share prices of all these are quoted by the newspapers mentioned above.

4 New regions for old

What are regions?

The newspaper cutting in figure 49 will confirm much of what you would already suspect from reading the previous chapter. It is true that this article names 'the South East', but you could take that to mean the central belt. You will notice that most of the poor areas in the article are in the outer area in figure 44.

1 Can you think of any reasons for the differences in house prices shown in figure 50? Have another look at figures 44 and 48.

In the last chapter, the words 'area', or 'part of Britain' were used to describe parts of the map. But the word 'region' might have been used instead. A region is an area that is in some way distinct from other areas.

In the case of figure 44 the shaded region is generally 'better off' than the areas outside it.

As with most regions, the boundary is difficult to draw. Really, there should not be an actual boundary, there should just be a gradual change from red to white. Since this would be very difficult to draw, we usually have to make do with a line of some kind. But when you see a map showing regions you should realise that usually the boundary lines do not mean a sharp break unless they are **administrative regions**, such as counties.

The regions in figure 51 are administrative. This means that the

Figure 50 (opposite) House prices in Britain

Rich – poor gap stays

The survey does show that on almost every measurable index of prosperity the division between rich and poor regions remains distinct.

Take first of all the consumer durables, the badges of our affluent society: more than half the households of the South East and East Anglia have at least one car, compared with about a third in Scotland. Indeed, 10 per cent of East Anglian households have two cars.

The South East also wins on central heating. About 47 per cent of homes in the region outside London, have it, double the figure for Scotland and treble that for Northern Ireland.

The North and Yorkshire win on washing machines, and national television usage has evened off at about 90 per cent throughout the country, but London telephone ownership, at about 50 per cent, is about twice that in the developing areas.

It is only when we come to the statistics of population density – with all that they imply in terms of congestion, housing costs and physical unpleasantness – that the figures start to tell the other side of the picture. Greater London's 4652 people to a square kilometre (under half a square mile) compares with no more than eight in the Highlands of Scotland.

More than 30 per cent of the UK's population lives in South East England, which covers 11 per cent of the land area, in spite of some small migration in the early seventies.

Figure 49

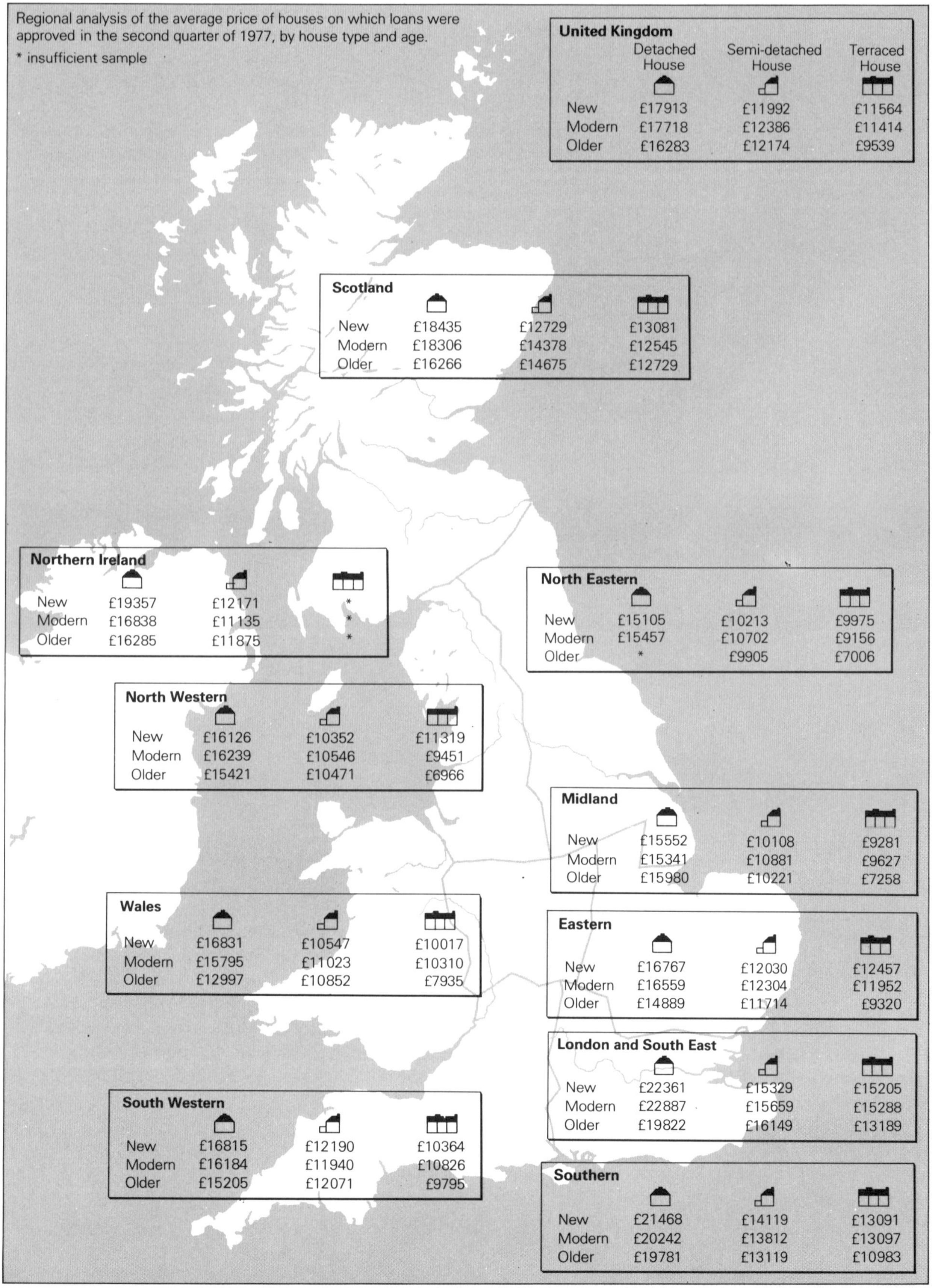

Regional analysis of the average price of houses on which loans were approved in the second quarter of 1977, by house type and age.

* insufficient sample

United Kingdom

	Detached House	Semi-detached House	Terraced House
New	£17913	£11992	£11564
Modern	£17718	£12386	£11414
Older	£16283	£12174	£9539

Scotland

	Detached	Semi-detached	Terraced
New	£18435	£12729	£13081
Modern	£18306	£14378	£12545
Older	£16266	£14675	£12729

Northern Ireland

	Detached	Semi-detached	Terraced
New	£19357	£12171	*
Modern	£16838	£11135	*
Older	£16285	£11875	*

North Eastern

	Detached	Semi-detached	Terraced
New	£15105	£10213	£9975
Modern	£15457	£10702	£9156
Older	*	£9905	£7006

North Western

	Detached	Semi-detached	Terraced
New	£16126	£10352	£11319
Modern	£16239	£10546	£9451
Older	£15421	£10471	£6966

Midland

	Detached	Semi-detached	Terraced
New	£15552	£10108	£9281
Modern	£15341	£10881	£9627
Older	£15980	£10221	£7258

Wales

	Detached	Semi-detached	Terraced
New	£16831	£10547	£10017
Modern	£15795	£11023	£10310
Older	£12997	£10852	£7935

Eastern

	Detached	Semi-detached	Terraced
New	£16767	£12030	£12457
Modern	£16559	£12304	£11952
Older	£14889	£11714	£9320

London and South East

	Detached	Semi-detached	Terraced
New	£22361	£15329	£15205
Modern	£22887	£15659	£15288
Older	£19822	£16149	£13189

South Western

	Detached	Semi-detached	Terraced
New	£16815	£12190	£10364
Modern	£16184	£11940	£10826
Older	£15205	£12071	£9795

Southern

	Detached	Semi-detached	Terraced
New	£21468	£14119	£13091
Modern	£20242	£13812	£13097
Older	£19781	£13119	£10983

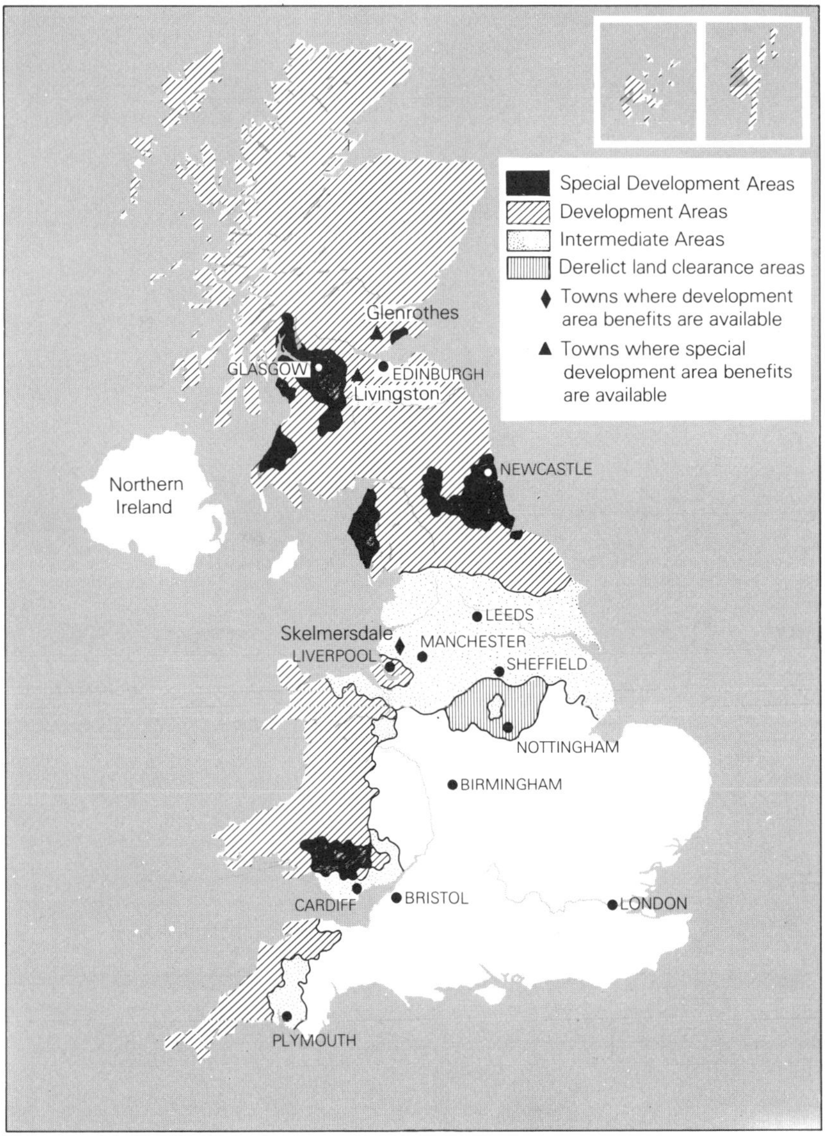

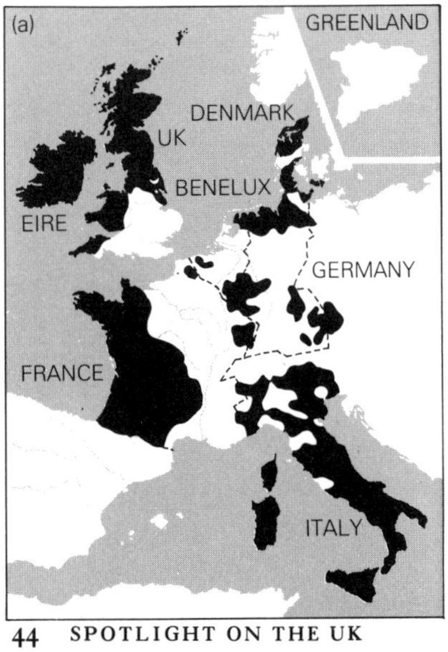

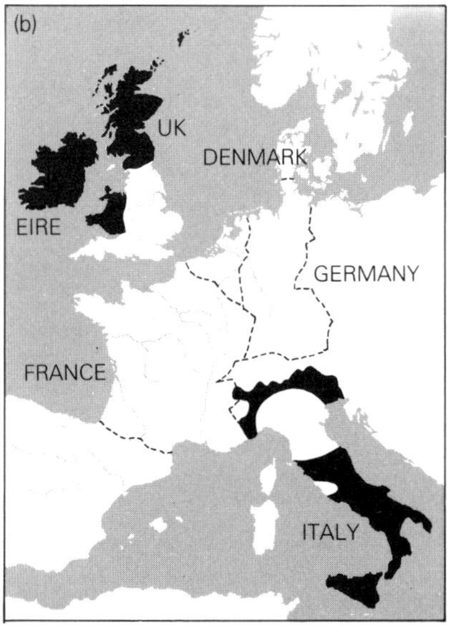

boundaries are lines which actually mean something. If you set up a factory in Northern Ireland you get more help than anywhere else. A Special Development Area gets the next amount of help, then a Development Area, then an Intermediate Area. The unshaded part of the map, which is very nearly the same as our 'rich' region in figure 44 gets no help. In fact it is in this area that Industrial Development Certificates are hard to come by.

Changing regions

2 At least three of the Special Development Areas in figure 51 were mentioned in chapter 2.

(a) Which are the three Special Development Areas?

(b) Which industries are found there?

Like most other aspects of human geography, regions change. In the last century, it was the coalfields and London that were the rich areas. Now, although London still makes the running, see where the coalfields are.

3 If in doubt about that last statement, compare figure 51 with figure 3(a).

The development areas map has also changed through time, according to the government in power and its attitude to economic problems. Figure 51 shows only the 1972 position in an arrangement which is continually changing.

There may be more changes in future. Since the UK joined the EEC, discussions have gone on about the help to be offered to the less

Figure 51 (above) Britain: Assisted areas in 1972
Figure 52 (left) (a) Europe's poor regions defined by European Commissioner Thomson
(b) Defined by German Chancellor Brandt

well-off parts of Europe. It is quite possible that European officials will decide that different parts of Britain need different amounts of help.

From chapter 1, you will realise that there is no such thing as a poor, or rich, region. There are simply areas which people *believe* to be rich or poor. You can see from figure 52 that the opinions of two influential people are very different, even though each was using the same measurement information.

Not only are the poor regions changing, so are the rich. And the problems are not all one-way. Most people live in the central belt, where most of the industries and the densest communication networks are. Thus it is also the most congested part of the UK. Moreover, it is becoming more congested all the time. Growth areas have problems too.

In the central area people have to travel further to work, often on overcrowded buses or trains, or in cars caught up in heavy traffic (see figure 53). Houses cost more as well, because there are more people wanting them and the land they are built on. The freight traffic on the roads creates noise nuisance, vibration (even threatening to shake down old buildings) and fumes which may be a health hazard.

At this stage, you could consider whether the two chapters on industry and these regional maps could help *you*.

4 When you leave school, what kind of work will you look for?

5 Will you always stay in your present home area, or will you hope to move away?

6 If you live in a development area, has your local authority an

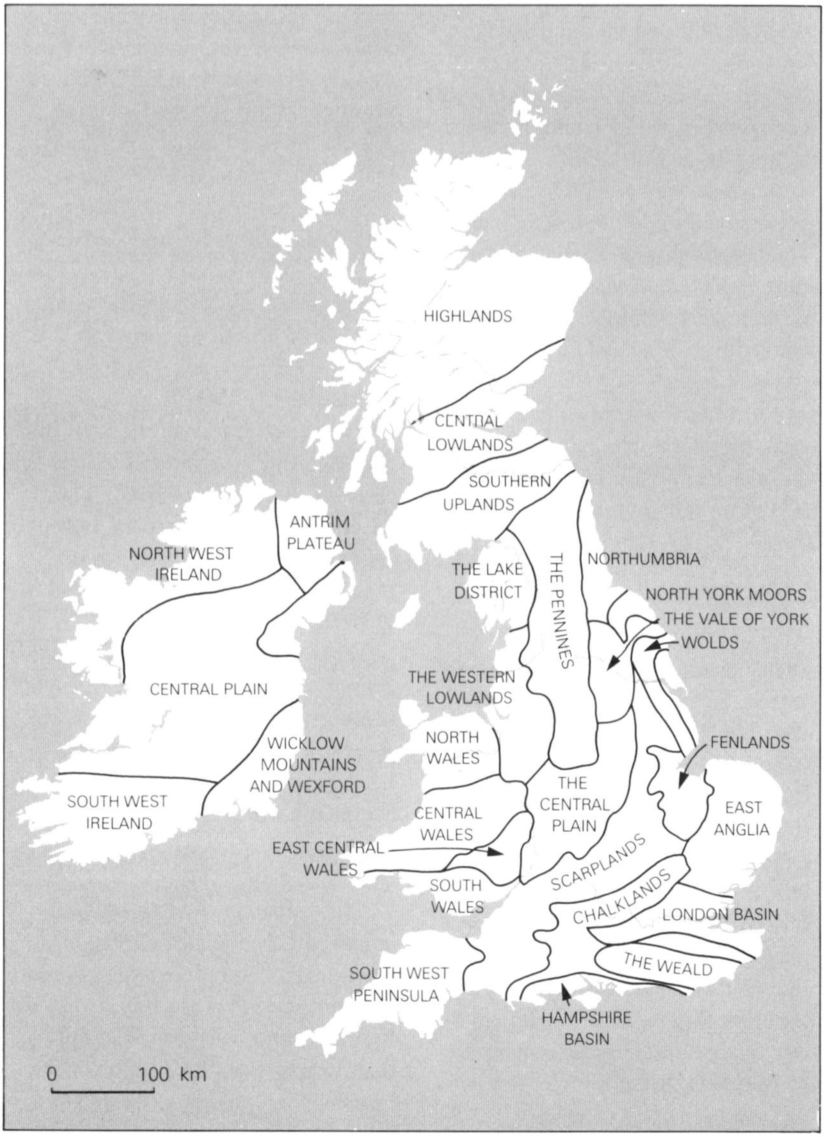

Figure 53 (above) *The long crawl home*
Figure 54 (below) *Physical regions of the British Isles*

industrial development scheme?

7 Which attracts you more, the possibility of higher income in a congested area, or a lower income but more open space nearby?

Physical regions

8 Why do you think figure 54 shows the whole of the British Isles and not just the UK?

9 What is a physical region?

10 The development area maps may be useful to you in reaching decisions. Do you think the map in figure 54 would be useful to you, or to anyone, for any purpose?

If you look at chapter 5, on leisure, you will find that there are references to the physical nature of the UK. The distribution of upland is important in understanding the location of National Parks, for example. It may also be useful in helping you to decide where to go on holiday. In chapter 12, on agriculture, it is obvious that an understanding of farming in the UK as a whole involves some knowledge of physical regions.

Knowing about this type of region, therefore, does have a purpose. Are there any other kinds which affect our daily lives?

City regions

There is one type of region which affects us much of the time – the city region. If you live in a city, much of your life is concerned with what is happening in it. You go to school there, buy food, clothes and luxuries there, you find entertainment there, and so on. But even if you do not live in a city, you still depend on cities for the same things. Although you may not buy food there, it is likely that you shop there fairly frequently for other things. It is probably in a city that you find entertainment, or go to school. Many people who live outside cities work in them. Cities also provide services, like gas, electricity and water, or hospitals and fire brigades. In addition, the people who live in a city often use the surrounding countryside for recreation, or they may get some of their food from it, so the connection between town and country is not all one-way.

If you look at the table in figure 55 bit by bit, you will find that it is not too difficult. First, Swansea CB means Swansea County Borough. This is just a little larger than the built-up area of the town. You can see where it is on figure 56. The two columns of figures are for male and female employees. A slight difficulty here is that all the figures are to the nearest ten. Every one therefore ends in 0 and this has been left off all the numbers. Thus the total working population of Swansea CB is 42 810 males and 27 050 females.

We are not really concerned here with looking at males and females separately, so, for each place, add the two together. Therefore, the total number of people who work in Swansea CB is 69 860. Of these, 57 870 also live there. Can you work out what percentage of the total this is? The other workers in the borough come from the places listed. The letters MB mean Municipal Borough; UD means Urban District. Both these are towns. RD means Rural District.

From the information we have, we cannot tell exactly where the people come from to work in Swansea, but at least we know the general area. Looking at the places where Swansea's workers come from gives us some idea of the size and shape of the city region.

Unfortunately, there is still one more difficulty. Perhaps some of the places named supply workers to more than one city. Perhaps, say, Bridgend supplies more workers to Cardiff than to Swansea. This problem can be solved by looking at the tables for

	Males	Females
Swansea C.B.		
Total working population	4281	2705
Area of usual residence		
SWANSEA C.B.	3370	2213
ELSEWHERE	911	492
Lancashire	6	—
Breconshire	21	8
Ystradgynlais R.D.	19	8
Carmarthenshire	127	42
Ammanford U.D.	10	2
Cwmamman U.D.	6	4
Llanelli M.B.	29	6
Carmarthen R.D.	6	1
Llandeilo R.D.	14	10
Llanelli R.D.	55	19
Glamorgan	709	431
Cardiff C.B.	12	4
Glyncorrwg U.D.	2	3
Llwchwr U.D.	169	137
Maesteg U.D.	5	1
Neath M.B.	34	17
Porthcawl U.D.	5	2
Port Talbot M.B.	45	39
Rhondda M.B.	7	—
Gower R.D.	211	119
Llantrisant and Llantwit Fardre R.D.	6	—
Neath R.D.	74	34
Penybont R.D.	12	3
Pontardawe R.D.	101	68
Monmouthshire	12	6
Newport C.B.	7	1
Pembrokeshire	5	1
Remainder of Great Britain	31	4

Figure 55 Extract from the work-place tables, 1961 census

Cardiff as well, but there is no space here for that.

You may think that looking at the number of people who work in a place is not a good measure of a city region. What about including all the people who use it for other purposes? To do this, you should find out where people come from who use the city for any purpose at all. Naturally, this is very difficult to do, if not impossible. One possible way is to go round nearby places and ask people if they have visited any town or city in the last week, or month. To ask every person would take too long, so a **sample** can be asked. If the sample is large enough, it is almost as good as asking everyone.

The map in figure 57 is not easy to

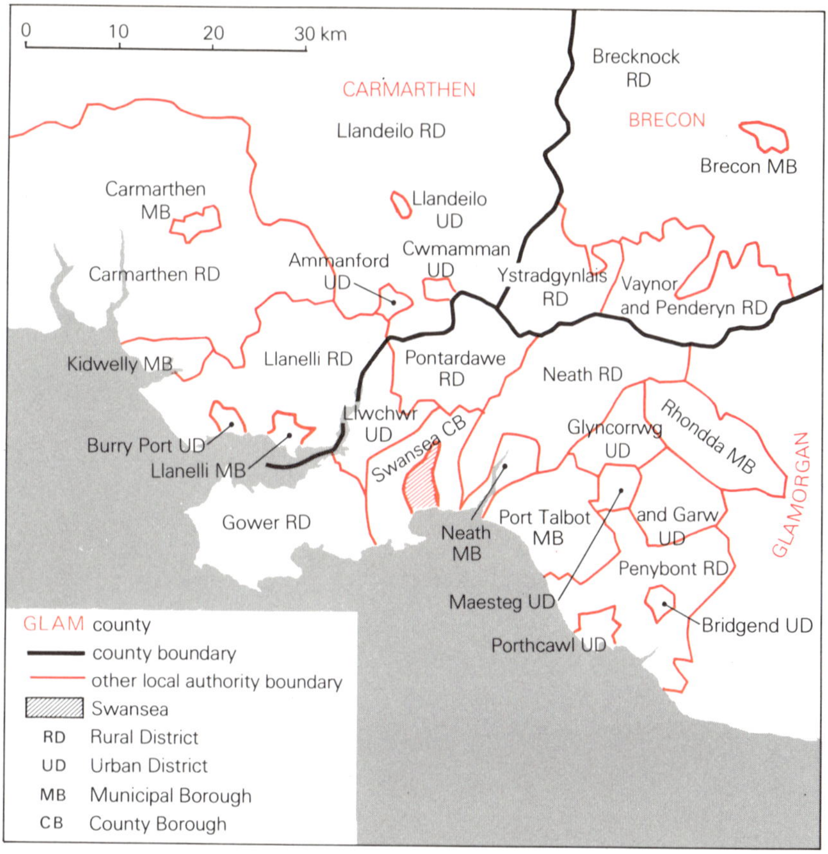

Figure 56 (right) Administrative areas around Swansea at the time of the 1961 census

Figure 57 (below) Frequency of visits to Swansea

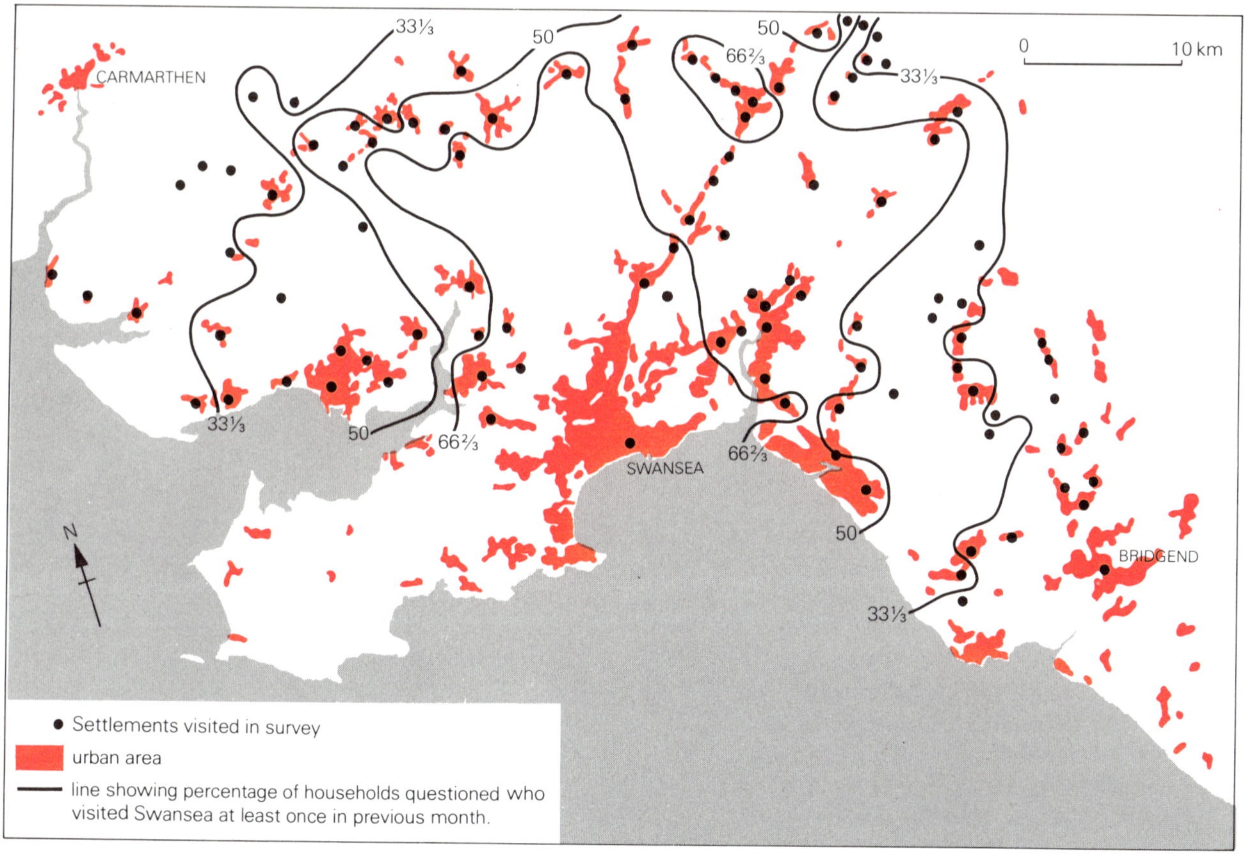

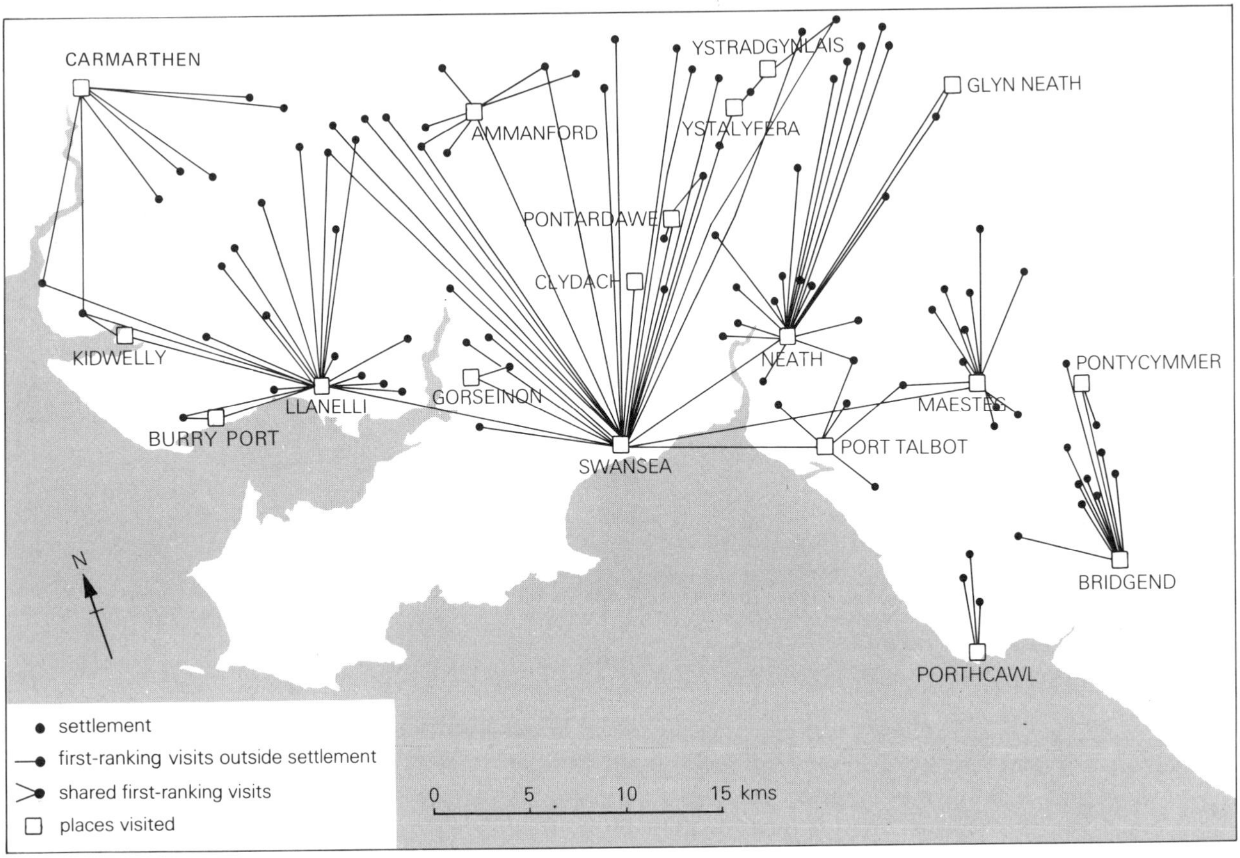
CARMARTHEN
AMMANFORD
YSTRADGYNLAIS
YSTALYFERA
GLYN NEATH
PONTARDAWE
CLYDACH
KIDWELLY
LLANELLI
BURRY PORT
GORSEINON
SWANSEA
NEATH
MAESTEG
PORT TALBOT
PONTYCYMMER
BRIDGEND
PORTHCAWL
N
settlement
first-ranking visits outside settlement
shared first-ranking visits
places visited
0 5 10 15 kms

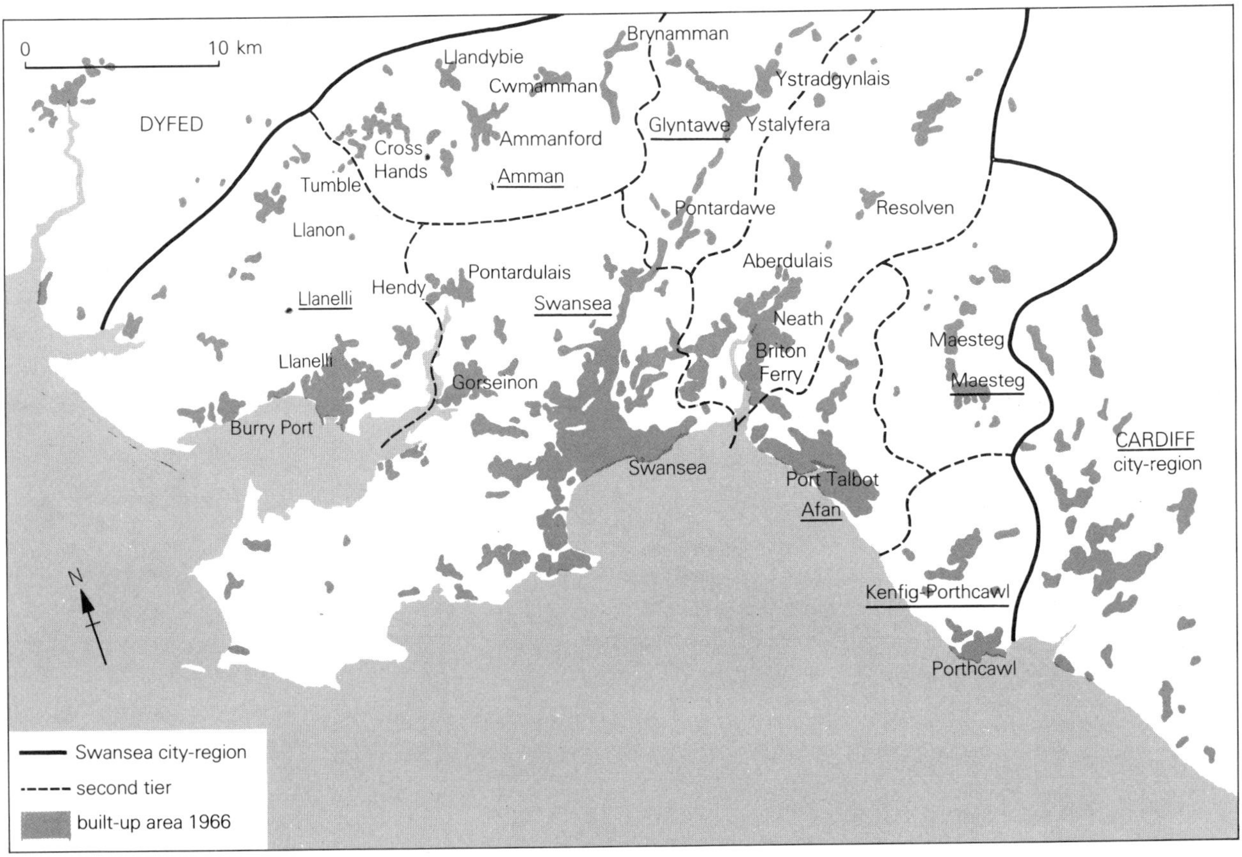
0 10 km
DYFED
Brynamman
Llandybie
Cwmamman
Ammanford
Cross Hands
Amman
Tumble
Llanon
Glyntawe
Ystradgynlais
Ystalyfera
Pontardawe
Resolven
Aberdulais
Pontardulais
Hendy
Llanelli
Swansea
Neath
Briton Ferry
Maesteg
Llanelli
Gorseinon
Burry Port
Swansea
Port Talbot
Afan
Maesteg
CARDIFF city-region
Kenfig-Porthcawl
Porthcawl
N
Swansea city-region
second tier
built-up area 1966

interpret and you may have to ask for help. However, once you understand it, it gives a fairly good idea of Swansea's city region. The map resulted from a survey of 3600 households conducted by Dr W. K. D. Davies of the University of Wales at Swansea.

Figure 58 shows the results of the survey mapped in another way.

In figure 58 each place in which people were questioned is shown as a dot. A line has then been drawn from the dot to the town or city which most people questioned in that place had visited in the previous two weeks.

You may have noticed that the words 'town' and 'city' have been used in this section without any explanation, but they are different. When *you* go shopping, for instance, if you wish to buy something frequently, you will not want to go too far for it, as the repeated journey would become too expensive. So perhaps you go to a smaller centre near at hand. If you live in the country this might mean that you would go to a village or small town, rather than a city. If you live in a city, it would mean that you would go to a suburban shopping centre, rather than to the city centre itself.

Figure 58 suggests that most people, though not all, made most of their visits over short distances to their nearest centre, such as Ammanford or Llanelli. Longer visits to Swansea, shown in figure 57, were made less frequently.

Figure 59 shows the local authority areas suggested by Dr Davies as a result of the survey. These authorities are 'two-tier'. This means that for important local government activities, such as town planning and rate collection and spending, Swansea would be the centre for the city region inside the thick line. Minor government activities would be centred in the small areas (shown by the thin line) which the survey showed to exist.

The boundaries of the smaller places all fit inside the boundary of the Swansea city region. This is called 'nesting'.

11 In your own words, explain what you think is meant by 'nesting'.

If you carried out a survey of farms and hamlets, you would probably find that 'village regions' also exist. These would nest inside the town regions. For instance, someone living in a hamlet might visit the nearest village daily to buy bread, cigarettes and newspapers. They might go to the nearest town once a fortnight to buy anything from a pair of babies' shoes to a record. Once every six weeks or so they might travel further still, perhaps to see a specialist at the hospital.

This seems very obvious. The point about it is that it shows there is some sort of order in the way people behave about visiting towns and cities. It does not, of course, mean that *everyone* behaves in the same way all the time. There are plenty of reasons why people visit centres further away than their nearest ones. They may think the journey is more picturesque; they may know the other centre better; the more distant place may have a better football team; and so on.

Even so, the study of Swansea has shown that most people, most of the time, behave in a way that has some order.

12 Why do you have to travel further to a Marks and Spencers than to a grocery shop?

13 Why do you have to travel to a secondary school?

14 Make a list of the various reasons why *you* travel into a town or city centre.

Local government regions

The study of Swansea has already been used to suggest new local authority areas. This is one way in which the study of city regions is very useful.

Before April 1974, the local government areas around Swansea were as shown in figure 56. The

Figure 58 (opposite, above) Visits in the Swansea area
Figure 59 (opposite, below) Proposed local authorities in the Swansea city region

Figure 60 Swansea area: present local authorities

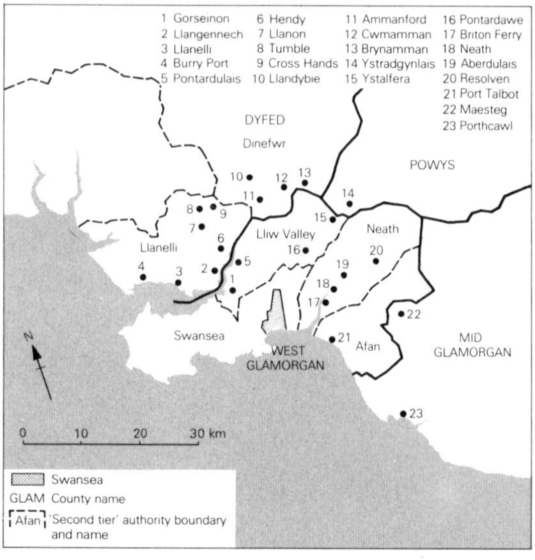

County Borough of Swansea was not much bigger than the built-up area of the city. The towns and villages around were in the counties of Glamorgan, Brecon and Carmarthen. This meant that each of these counties (a county borough worked like a county) had separate planning schemes, they looked after their own roads, their education systems were separate, and they each collected their own rates to pay for these things.

15 Can you think of any other local government services which are paid for out of the rates?

This meant that although the people from the whole area used Swansea regularly, only the people and businesses of the city paid rates for services which the visitors used.

Would it not be better to have a completely new county, like the city region in figure 59? In this way, all the people who used Swansea would be governed from the same place. Future planning and transport arrangements would be done by a single authority, instead of the four involved previously. If city regions exist, would it not be better if they also became administrative regions, rather than just something cooked up in geography books?

As we have seen, Dr Davies suggested that the Swansea city region, which his survey showed to exist, should become just such a new county. Well, he almost got his wish, but not quite.

The Local Government Act of 1972 decided that there should be new counties, starting in April 1974. As far as possible, these new counties were to be city regions, at least in the most urbanised areas of Britain. In deciding the new county boundaries,

Figure 61 New counties of England and Wales, and new regions of Scotland

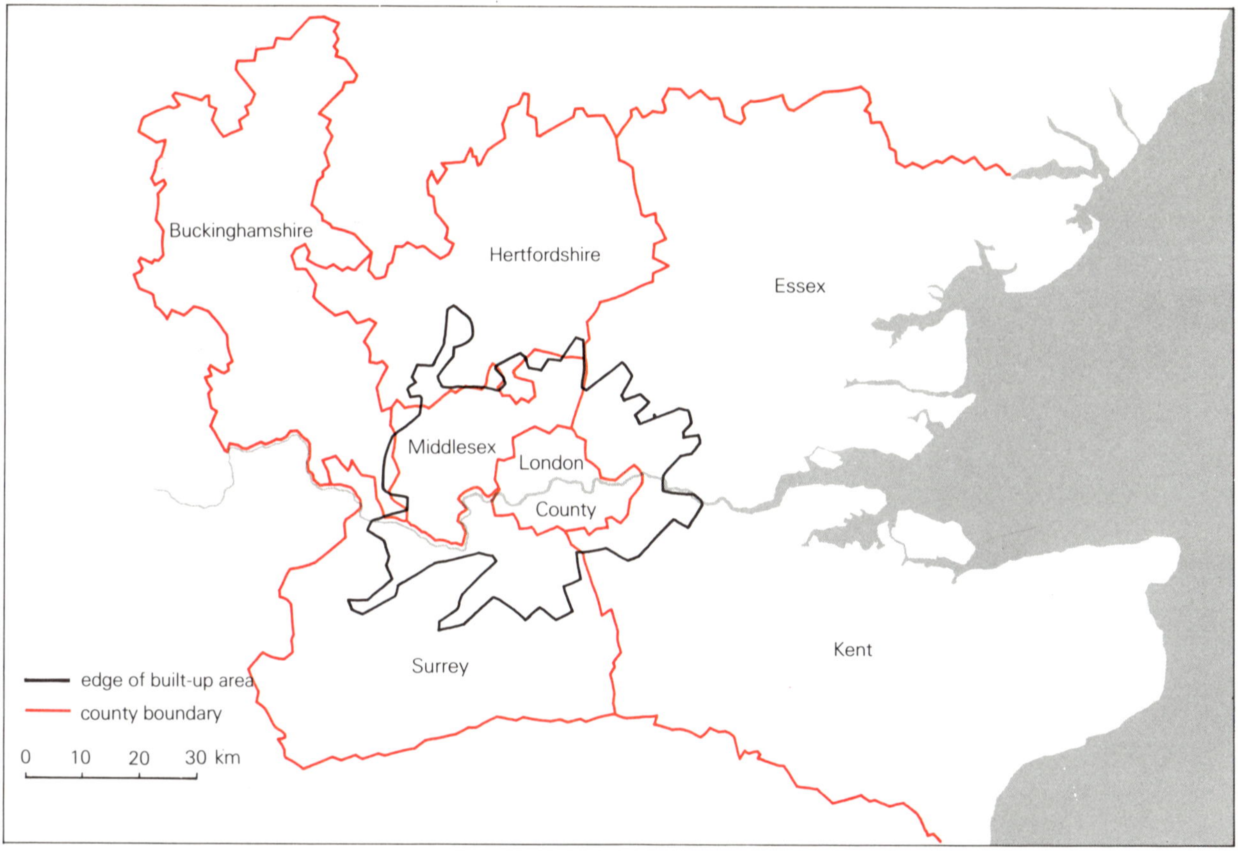

Figure 62 London area: administrative regions before 1965

however, there were not always surveys like Dr Davies' to help. The wishes of the existing counties and of the local people also had to be taken into account. This meant that the new counties were not always the city regions that geographers would have devised.

Compare figure 59 with figure 60.

16 What is the name of the county that includes Swansea?

17 Which places that were in Dr Davies' suggested Swansea county are now in the county of Dyfed?

18 Which is in Mid-Glamorgan?

19 Which is in Powys?

Although new counties were needed in the Swansea area, the need was probably even greater in other parts of Britain. The great conurbations (see page 18) had started as separate towns, but have grown together into enormous built-up areas. But they still kept their original local governments. This was obviously ridiculous. Although the conurbations had many different centres, they were really large super-cities. Every part was dependent on every other part. Every day there were huge flows of people, goods, letters, telephone calls between the different 'towns' of the conurbations. Thus it was essential that planning, transportation and other major decisions should be made by an authority covering the whole area. The difficulties of deciding which places to include in the new 'metropolitan counties' were enormous. This is one of the reasons why we used the simpler case of Swansea as an example.

Figure 61 shows all the new counties of England and Wales and the regions of Scotland.

London's region

No mention has been made so far of London. This is because, as in many other ways, London is a special case. Before 1965 the administrative regions near London were as shown in figure 62.

20 Using figure 62 make a list of the counties which contained part of London's built-up area before 1965.

21 Which county boroughs were included within the built-up area?

In 1965 the Greater London Council (figure 63) was created, to take control of overall planning and transportation for the whole area. The area was so big, however, and contained so many people, that each of the thirty-two London boroughs had a lot of control over its own affairs. It has to be remembered that each London borough contains as many people as a fair-sized city.

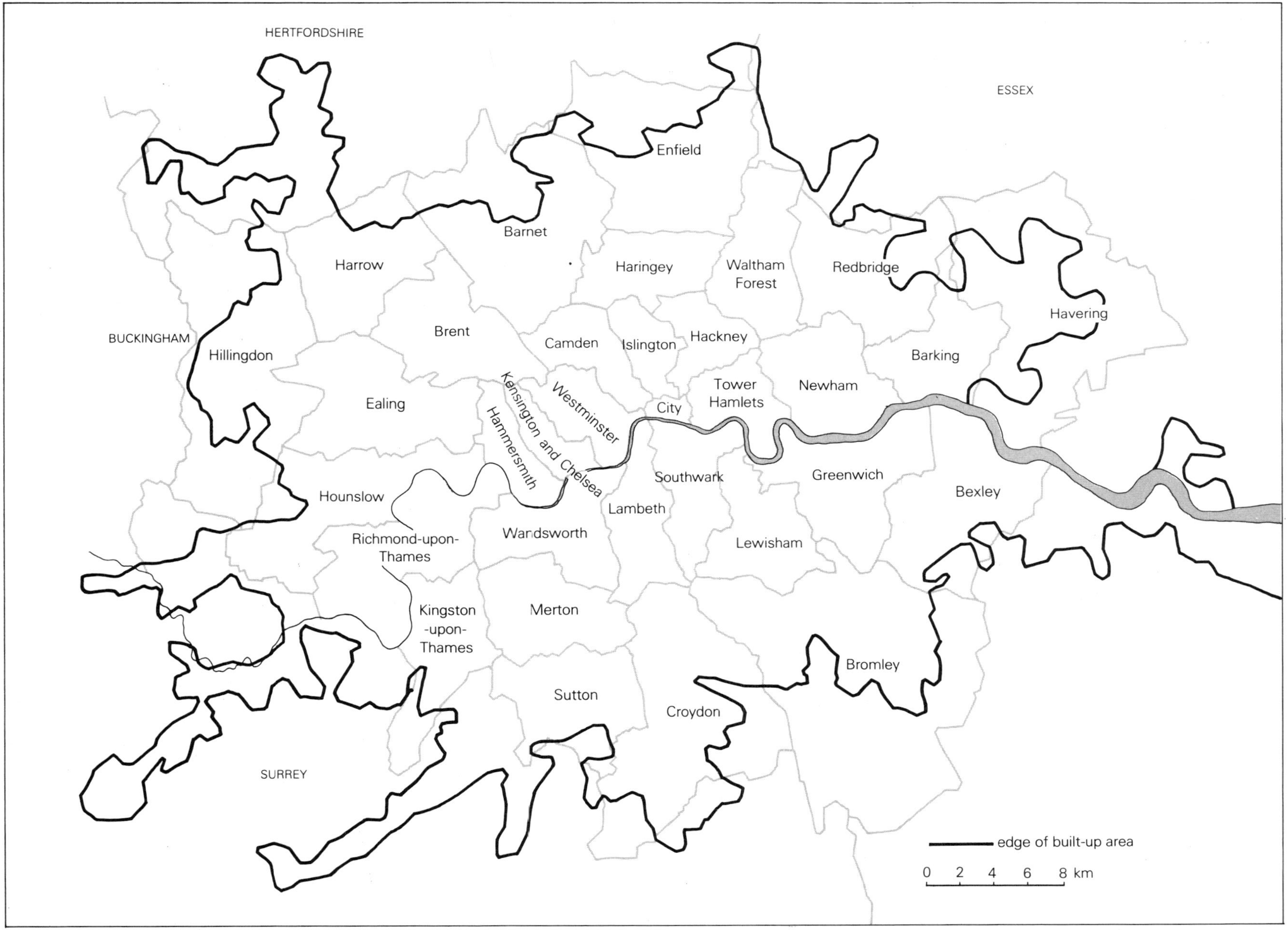
HERTFORDSHIRE
ESSEX
BUCKINGHAM
SURREY
Enfield
Barnet
Harrow
Haringey
Waltham Forest
Redbridge
Havering
Brent
Camden
Islington
Hackney
Barking
Hillingdon
Ealing
Kensington and Chelsea
Hammersmith
Westminster
City
Tower Hamlets
Newham
Southwark
Greenwich
Bexley
Hounslow
Lambeth
Richmond-upon-Thames
Wardsworth
Lewisham
Kingston -upon-Thames
Merton
Bromley
Sutton
Croydon
edge of built-up area
0 2 4 6 8 km

Figure 63 (opposite) *Greater London*

Although the creation of the GLC was intended to solve problems the change was not entirely successful. Many people believe that the GLC is too big. They say it is not able to handle the huge problems of the whole area. In addition, it is claimed that it is too impersonal. *Local* government should be really local not a remote and massive organisation such as the GLC.

In this chapter, we have found out what regions are and how geographers describe them. We have seen how geographers' ideas of a city region have been put to use in devising new counties. A new bit of British history has been made, and geographers have played an important part in it.

PROJECTS

1 The easiest type of region to measure for yourself is your own school's catchment area. You could find this out from the school office. If you wanted to try out a research method, you could ask a sample, say 20 per cent of the school students, where they live. Each place could be marked as a dot on tracing paper placed over a one-inch or 1:50000 map. The outermost dots could then be joined up to give an idea of the catchment area.

If you wanted to see how well your sampling had worked, you could compare your results with the official area.

If your sampling did not work well, can you think of any reasons why this should be so?

2 If you go to school in a small town, your school's catchment area will be a guide to the town's region.

However, there are other ways of trying to find out your town or city region.

(i) Find out the area within which a large store in the town centre delivers its goods.
(ii) If there is a local newspaper, find out over which area it is sold, or where its different editions are sold.
(iii) Find out the area from which the hospital draws its patients.

These are just three suggestions for looking at your own city region. You will find that each has a different boundary. Thus the boundary of the city region will be difficult to draw. However, we have already seen that boundaries to regions are not very satisfactory.

You may be able to think of other ways of defining your own city region. If you do, you will probably find each one has a different boundary.

3 If you live in a very big city, the previous exercise may be too difficult. However, you could still look at the sales area for the principal local papers. For example, in London, find out where the *Evening Standard* and *Evening News* are sold. It is a huge area! Similarly, the *Birmingham Post* might be a good indicator for that conurbation.

4 Look up the workplace tables of the census for your town or city. It will be necessary to look up other towns in the area to see whether people go to them as well as to your town. If, say, a village sends workers to more than one town, you will have to work out a way of deciding whose region it is in.

5 Playtime

Space for playtime

1 Can you name five **leisure** activities which take place outside the home from figure 64?

They are very different kinds of entertainment, but they have one thing in common; they need a *place* in which they can happen. This place needs to occupy a certain amount of **space**. The extent of the space needed is different in each case. A disco may need only the size of one room; a cinema or bingo hall would need more, and a football pitch more still. For hill-walking to be enjoyable, most people would agree that a space of several hundred square kilometres is necessary.

2 Here is a list of dimensions of various sports areas. See if you can fit the sizes to the correct sport.
(i) 100–110 metres × 70–80 metres; (ii) 26 metres × 12 metres; (iii) 100 metres × 55–60 metres; (iv) 28 metres × 15 metres; (v) 20–30 metres × 8–20 metres; (vi) 11 metres × 7 metres
(a) squash court; (b) basketball court; (c) football pitch (international); (d) water polo; (e) tennis court; (f) hockey pitch.
If you do not know the answers, have a guess. Then check your guesses with an encyclopedia or sports book.

One of the difficulties with space is that in a small, crowded island like ours, it is very valuable. Different people need space for many different purposes, and the more people want a thing, the more expensive it will be. In towns space is very expensive

MECCA SOCIAL CLUB
AFTERNOON
EVENING
YORK'S MOST LUXURIOUS
Social Club
RIALTO
MECCA Bingo
COME & PLAY
THE ELECTRONIC WAY
THE MECCA WAY

Figure 64 (opposite)

indeed and therefore the sorts of leisure pursuits to be found there need to pack a lot of people into a small area in order to pay the rent. If a large space is needed near a town centre for only a few people, then either the users have to pay a lot of money, or the cost has to be met by someone else. A typical example of this is a local club football pitch, where twenty-two players need a space much bigger than a cinema holding several hundred paying customers. In cases like these, the local authority often provides the spaces and allows people to use them fairly cheaply.

Not everyone may believe that the council is right in using valuable land for recreation. Suppose the town has a housing shortage. Should not the land be built on to provide houses for the homeless? Or perhaps the land should be rented for office block use. The rent and rates from the offices would produce enormous income for the authority. This would mean that for the residents, better social services could be provided for lower rates.

Local authorities have to think about such matters when deciding how best to use the land available, but they *all* provide some outdoor leisure space. *Outdoor* use is named specially here because these are the uses where a few people need a lot of land. This is called **space-extensive** use.

One of the reasons why leisure space is provided is because it is thought to be important for the mental and physical health of ordinary people. Just as the homeless may get illnesses brought on by cold and exposure, so people without recreation are thought likely to suffer, though in less obvious ways. This is what the Greater London Development Plan has to say about it:

Figure 65

'. . . but open space planning has suffered too often from needs for economy and the council considers that a fair distribution is important to the life of London as a whole. At the same time, local authorities need to be vigilant in preserving what open space already exists. Proposals to use recreational open space for other developments need to be treated with special caution.'

3 Do the GLC planners think that using land for recreation is less important than other uses?

4 What has the use of open space suffered from in the past?

5 Do you think it possible that open land could be used up without anyone noticing until it was too late?

6 What does the GLC urge local authorities to do to prevent this happening?

So far, in discussing leisure, the only types mentioned have been sport and entertainment. But in providing open spaces, authorities have to remember the people who just wish to relax and look at the view. A survey of people's reasons for visiting parks produced the following figures: 86% sitting, walking, enjoying the view; 12% children's games; 6% sport; 3% entertainment.

7 These numbers are percentages. Why do you think they add up to more than 100?

8 In figure 65, which of the following types of people can be seen?

(a) Mothers or fathers with young children;
(b) Retired people;
(c) Workers at lunchtime;
(d) Team games players;
(e) August bank holiday family outing.

Naturally, from a picture you cannot be certain, but you will be able to make a good guess.

9 Which types of people would be most likely to visit a town park

(a) daily?
(b) weekly?
(c) occasionally?

Question 8 should give you some clues to answer this one.

Of course, you do not have to go to parks to look at the view. At weekends or during holidays car owners often drive quite a long way. They are not looking for a park. All they are seeking is a bit of open country away from the noise and bustle of the town.

The National Parks

Even outside the towns, Britain does

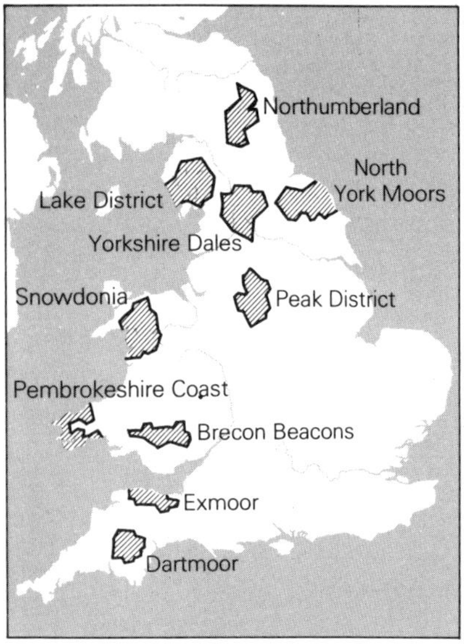

Figure 66

not have very much open space available for leisure. The land is needed for farming; it may also be needed for some types of mining or industry. Near towns and villages, it may be needed for housing. Because of these different needs, some of the most beautiful parts of the country have been turned into 'national parks'. The idea is to let the land continue to be used for farming, or other 'rural' activities, but to prevent it being spoilt by industry or by towns spreading into it. At the same time, it is hoped to reach agreement with farmers and other landowners to allow visitors on to their land and to make provision for the visitors, such as footpaths, picnic areas and information centres so that the agricultural land is not damaged.

Large areas of land like National Parks are naturally going to attract other people besides casual visitors. Since the areas are often uplands, hill-walking is popular. Climbing, too, is an attraction, and where there are lakes, boating and water sports are possible.

10 (a) From an atlas, trace a map of Britain and shade in all the land over 100 metres.

Figure 67 (above, right) The Winnats, near Castleton, Derbyshire
Figure 68 (below) This picture taken in 1932 is not very clear, but it is the same area as figure 67, and if you look carefully, you can pick out the same rock formations

(b) Compare your map with figure 66. How many of the National Parks have most of their land over 100 metres?

(c) Can you think of any reasons for your answer?

With many different users, it is hardly surprising that not everyone agrees about how a National Park should work. If, in addition to the uses intended for the park, there is an important mineral to be mined or some other attraction for industry, then perhaps this should be allowed as well.

Suppose you live in a National Park and happen to be unemployed. Your best friend has a good job and is a keen rock climber. A rock outcrop could be quarried and provide you with a job, but the same rocks provide your friend, and hundreds of others from the cities, with weekend relaxation. What do you think should happen to the rocks?

Figure 69 Solid geology of the North of England

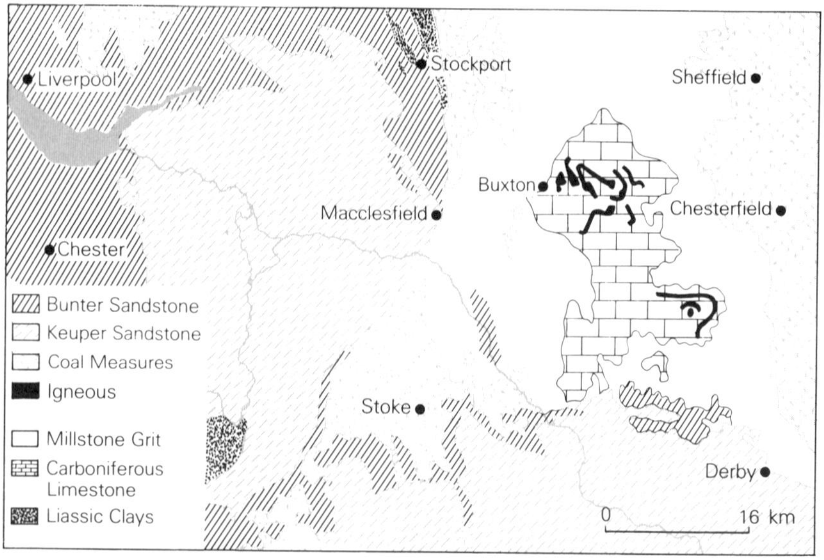

11 Figure 67 was taken in the Peak District National Park.

(a) What recreational activities might be popular in an area like this?

(b) What disagreements might there be between people who think the land should be used differently?

Figure 68 shows that demonstrations are not new. The people were demanding a legal right to walk on uncultivated mountain and moor land. The following extract from a Peak Park brochure explains why.

'Before the National Park was established, if you went up on most of the high moorlands such as Kinder Scout and Bleaklow you were trespassing – because the moors were preserved for shooting. Landowners were concerned about the effect of large-scale public access but, with crowded cities nearby, there was strong demand for public access. In 1932 there was a mass trespass of ramblers and walkers on Kinder Scout. . . . You can now have access to seventy-three square miles of open country though the land still remains privately owned. Visitors must observe bye-laws. . . . Walking, shooting and upland farming have been reconciled.'

12 What do you think the bye-laws mentioned at the end of the extract are likely to be about?

13 If you do not know them, look up in a dictionary the words 'reconcile' and 'reconciliation'.

Reconciliation is not always easy. The Winnats Pass is carved out of Carboniferous Limestone, which is a very valuable mineral. It is even more valuable when it is found near coal, because both are used in making steel, and near shale, because both limestone and shale are used in making cement.

Figure 70 (a) (above, centre) Tunstead quarry (b) (right) Hope cement works

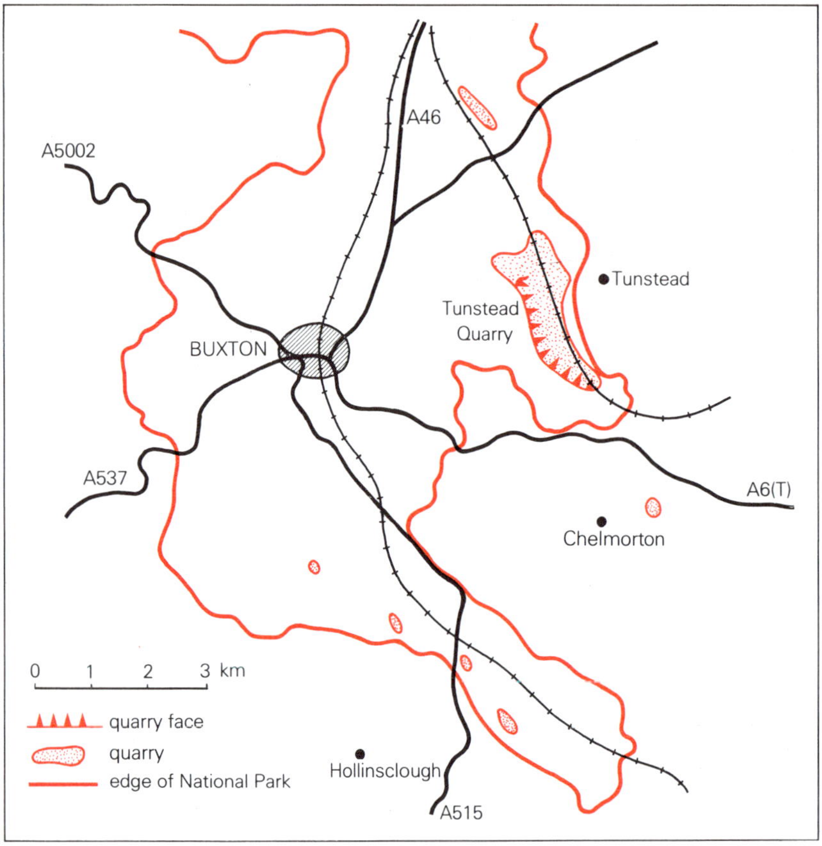

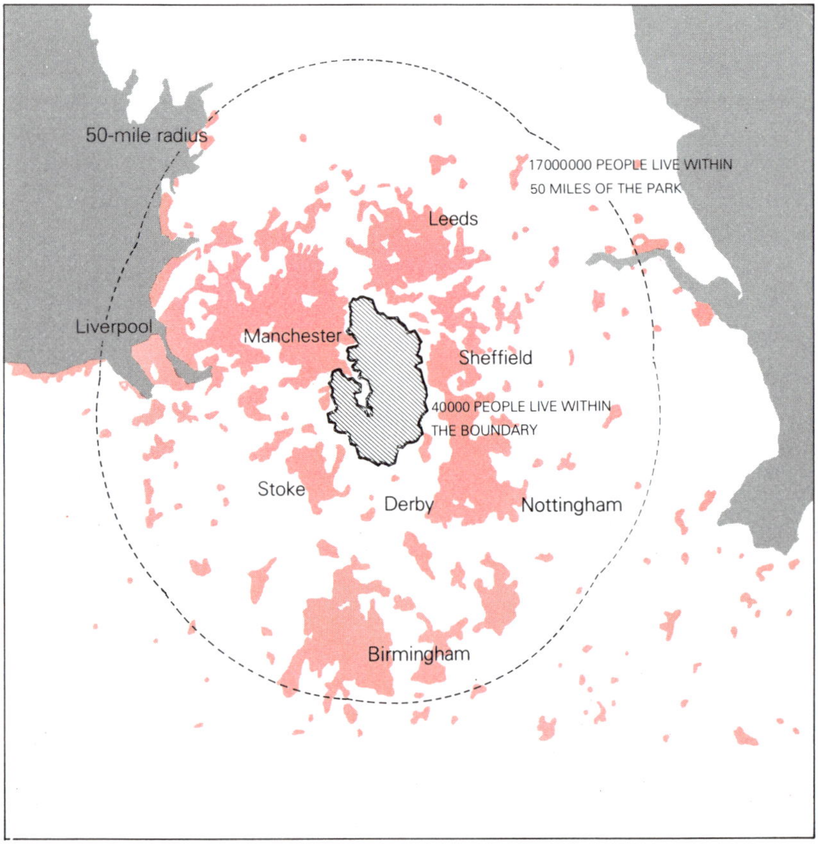

Figure 71
Figure 72 (below) The Peak Park in its urban setting

14 Look at figure 69. Find the area of Carboniferous Limestone around Castleton.

15 Can you find a place connected with steelmaking on a nearby coal-field?

The huge limestone quarry in figure 70 is outside the park, though right on the border. It is marked on figure 71. The cement works is right in the middle of the park. It has its own large limestone and shale quarries.

16 Look at figure 71.
(a) How many times can you find the word 'quarry'?
(b) How many are inside the National Park, how many outside?
(c) Why do you think there is a big 'bay' in the boundary of the park in this area?
(d) How long, from north to south, is the Tunstead quarry?
(e) How wide is it, from the railway to the quarry face?

You may think that such workings are out of place within, or on the edge of, a National Park. But in addition to the importance of the limestone, quarrying provides one of the main types of work for the people in the park. If you take away the quarries, you take away thousands of jobs. What would you do?

This is not the end of the conflicts facing park planners. Another difficulty is created by the users themselves. Figure 72 gives some idea of how the problem arises.

At the time of the demonstration for access to the open lands, most of the users were walkers or climbers, who probably travelled to the area by train. Although they held a mass trespass, there were not really very many of them.

17 Since the Second World War,

Figure 73

what do you think has happened to:

(a) the population around the park?

(b) the amount of time which people have for leisure?

(c) the number of people with cars?

18 What has happened over the last few years which makes the parks easier to get to for people with cars, particularly those who live more than eighty kilometres away?

Nowadays, the 40 000 residents of the park have 16 000 000 (sixteen million!) visitors each year. Most of these are not climbers or ramblers, but families out for an afternoon's drive. So many cars can cause jams on country roads and bring fumes and unsightliness to beautiful places. Ramblers, climbers, pony trekkers and others trying to 'get away from it all' would perhaps like to see cars banned. They may find support among the farmers who also suffer from the citizens shown in figure 73.

But the parks are for everyone, not just a selected few, so what is to be done about it? There are no easy answers, but at least some improvements would be gained by providing

Figure 74 Tideswell Dale (a) before (b) after

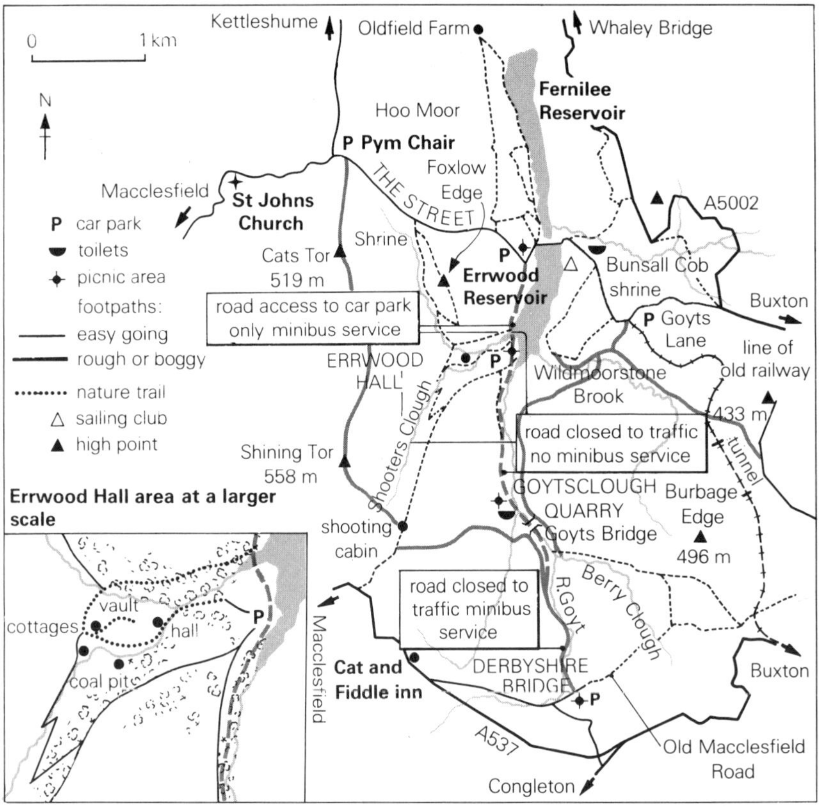

The scheme

In the summers of 1970 and 1971 the Countryside Commission and the Peak Park Planning Board ran a week-end traffic experiment in the Goyt Valley, to see whether 'park and ride' could cope with traffic problems at rural beauty spots. In general the experiment worked very well and surveys showed that closing the valley roads was very popular with visitors, who were able to abandon their cars and enjoy walking along the traffic-free roads and way-marked paths, or riding on the mini-buses.

Removing most of the traffic lent a general air of peacefulness to the area; visual intrusion by parked vehicles was removed, traffic noise reduced, and narrow roads formerly badly congested with traffic were released for almost exclusive use by pedestrians.

Since 1971 the Peak Park Planning Board has been responsible for traffic management in the valley, and the extent of traffic restrictions has been reduced as a result of experience gained from the experiment.

The scheme is an exercise in managing the countryside for the benefit of all concerned; the motorists, cyclists, horse-riders and walkers who come here for enjoyment, as well as for the farmers, foresters and game-keepers who make a living in the area. Visitors are not banned from the valley. On the contrary, facilities have been provided so that the beautiful surroundings can be enjoyed to the full, without the unpleasantness and congestion created by the motor-car.

Timing

The scheme operates on Sundays and Bank Holiday Mondays from the last Sunday in May to the last Sunday in September. The restrictions are in force between 10.30 a.m. and 6.30 p.m.

Car parks

No cars are permitted to use the main valley road between Errwood Hall and Derbyshire Bridge when the traffic scheme is running. Visitors arriving between the hours of 10.30 a.m. and 6.30 p.m. must park at the edge of the motor-less zone at the free car parks at The Street and Derbyshire Bridge, and then either walk into the valley or ride on a mini-bus.

At the northern end of the valley a limited number of cars are allowed access from The Street to Errwood Hall car park for a charge of 20p per car, motorcycles 10p. Each car park is supervised by a warden with powers to direct traffic. He will also give information and advice to visitors.

Picnic areas

There are picnic areas adjoining each of the three car parks, and also at Goyts-clough Quarry. Please use the litter baskets which have been provided at these sites.

Mini-buses

A mini-bus service carries visitors from the outer car parks at The Street and Derbyshire Bridge to the traffic-free area for a small charge per head.

Exceptions

Motor access through the barriers at Derbyshire Bridge and Errwood Hall is only permitted to those who require access to land in the valley, i.e., those carrying permits, as well as the essential services such as police, ambulance and fire brigade. These exceptions are only few in number.

Pedal cyclists, walkers and pony trekkers may use all the roads in the valley. The wardens will exercise discretion in letting through private cars carrying very old, infirm or disabled people, who are unable to use the mini-bus service; from The Street to Errwood Hall and from Derbyshire Bridge to Goytsclough Quarry.

Figure 75 The Goyt Valley traffic scheme

more parking and picnic places, like the one in figure 74 which has made use of an abandoned quarry.

Another idea, tried with success in the Peak Park, is to ban cars from particularly beautiful places, but to let people see them by means of free minibus rides from car parks at the edge of the area.

Look at figure 75 then answer the following questions.

19 What is the reason for the scheme?

20 How are people able to get into the Goyt Valley if they do not wish to walk?

21 Make a list of the facilities provided for visitors.

22 Make a list of the things you can see on the map which might be an attraction for visitors.

23 (a) Do you think this scheme is a good idea?

(b) Do you think cars should be let into the valley?

24 Can you think of any ways to improve the scheme?

Before leaving National Parks, it is necessary to look at another one, where the problems are similar, but different. Look at figure 76.

Most of the Pembrokeshire Coast National Park, is, as its name suggests, by the sea.

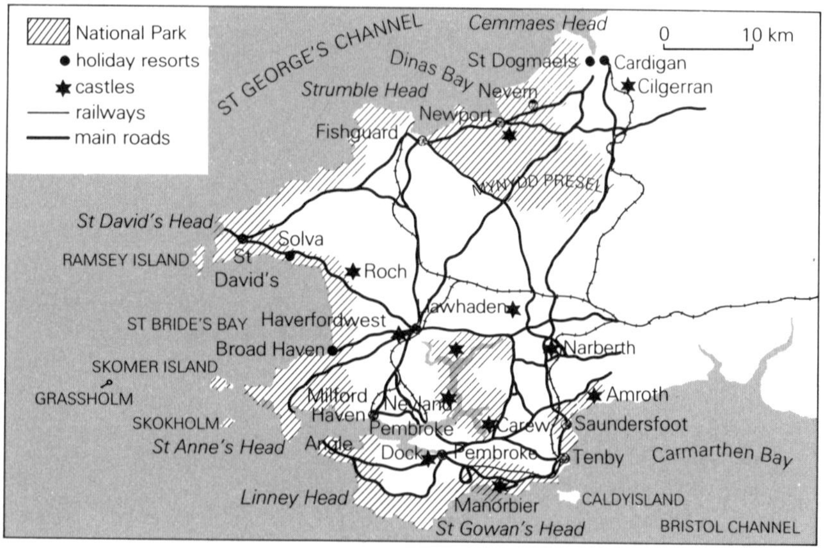

Figure 76 The Pembrokeshire Coast National Park

Figure 77 shows a harbour in the National Park.

25 How deep is the deepest water shown on the map in figure 77?

26 How far inland (from level with St Anne's Head) does the deep water extend?

Because of the deep water a long way inland, and because of the right-angle bend near the mouth, Milford Haven is a very sheltered harbour for big ships.

27 Which ships these days need the greatest depth of water?

28 Which of these ships are likely to use Milford Haven, a long way from the nearest big cities?

Figure 78 shows one of these ships near the entrance to the Haven. Because this harbour is ideal for big tankers, the major oil companies use it and have built refineries. There are four large refineries and an oil storage depot.

Figure 79 shows part of an oil refinery. See how many narrow vertical structures there are.

Figure 77 Milford Haven

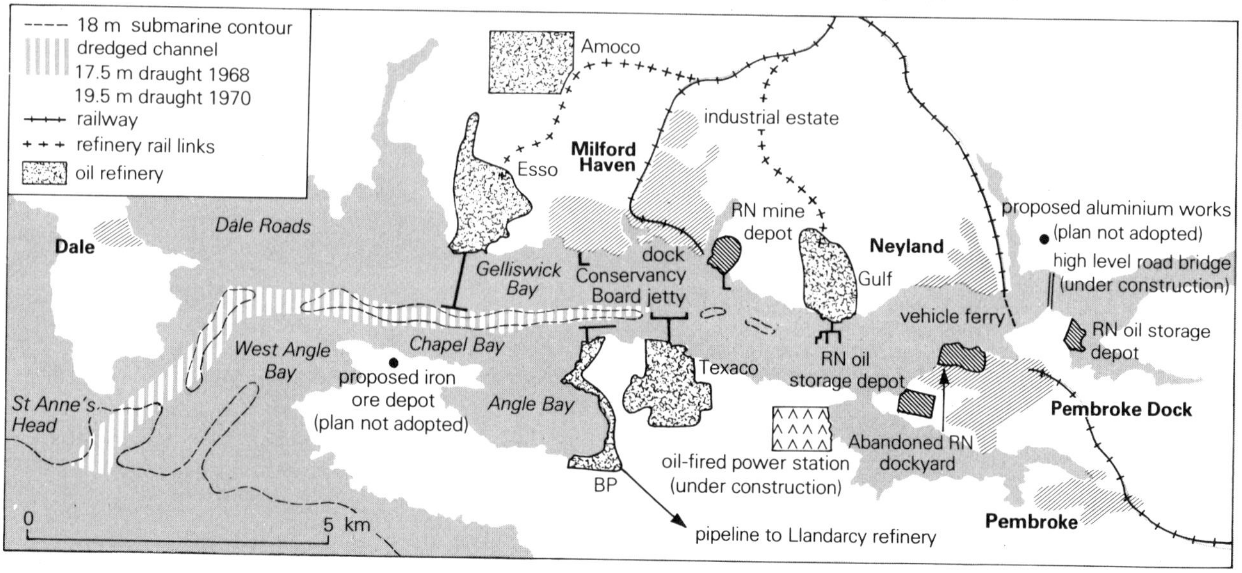

29 Look again at figure 77. Do you notice anything about the horizon?

If you had to design structures which were most noticeable in an area with an almost dead flat horizon in every direction, which shapes would you choose? . . . Yet this is a National Park! Milford Haven could be a splendid centre for recreation. Indeed, there are several sailing clubs, but much of the space is taken up by oil jetties and many of the shores are polluted by oil.

It may seem a pity that this magnificent landscape is spoilt in this way, yet in the area of Milford Haven, unemployment is much higher than the national average. The oil refineries help to provide work. Again, oil is vital to the nation. Wherever they are put, refineries are going to be a nuisance to someone.

Space for holidays

Last century, it is said, the weekend had not been invented. This century, the weekend, the week's and the fortnight's holiday have become the basis of a multi-million pound industry. More people are having longer holidays than ever before. Far more people are travelling by car, and more are going abroad. But for those who take their holidays in Britain, where are the most attractive places? What are holidaymakers looking for? Traditionally, the main ingredients are sunshine, sand and somewhere comfortable to stay.

The motor car, again, is altering things. Many families still have a traditional style of holiday, but others are taking to caravanning, camping or touring.

30 What do you think is meant by a 'traditional style of holiday'?

Even if you stay in one centre, you can make day visits if you have a car. Perhaps therefore, in addition to sun and sand, more holidaymakers are now demanding beautiful scenery, or historic monuments, or some other attraction as an alternative to spending the day on the beach, or wandering around the town if wet.

Figure 78 (above)
Figure 79 (below) The Waterston refinery, Pembrokeshire

31 The South West of England is a favourite area for holidays. Figure 80 shows some of its attractions. How many different things can you find that might attract holidaymakers? There are at least twenty, but perhaps you can find more still.

The South West, then, is very attractive. But, as we saw in the Peak Park, popularity brings its own problems.

Although it is not as near to the big population centres as the Peak Park, so many people wish to visit the South West, that scenes like the one in figure 81 are normal in the holiday season. The roads of the area were mostly developed for use by farm traffic. They cannot cope with the motor car boom. Any attractive piece of scenery, or even a moderate beach, if near a road, becomes

Figure 80 (opposite)

BM 40

Figure 81

choked with parked cars.

As usual, solutions are not easy. If one provides motorways and huge car parks the very things people have come to see are destroyed. Even as it is, the completion of the M5 motorway means that the South West is now within a few hours' drive of both London and the industrial Midlands.

Another problem concerns the work of the local people. During the summer, most of the inhabitants are connected with the holiday trade. Although farmers continue to farm, many of them get some income from renting accommodation, or from campsites and car parks on their land. A great many people rely entirely on the tourist industry, hoping to make enough money in the summer to tide them over the quieter winter months.

Thus it is vital, both for the holidaymaker and for the local people, that the area is maintained in an attractive way. But how can this be done?

32 Now that you know something of attempts in the Peak Park to solve the problems caused by large numbers of visitors, what would you do in South West England? Either make a list of ideas, or better, try to devise a complete scheme (perhaps along the lines of the Goyt Valley) for one small area in Cornwall or Devon.

PROJECTS

1 Look at a copy of OS 1:50000 sheet 180 (Barnstaple and Ilfracombe). Imagine you have decided to go on holiday in the area bounded by eastings 40 and 55 and northings 37 and 49. You have no car.

(a) Make a list of the features you most like on holiday (beaches, rock coast, plenty of people, emptiness, scenery etc.)
(b) Assume that you wish to stay in a town or village. By looking at the map, decide which place is likely to be the best for your holiday.

Of course, the map does not give you all the information you would need, but it gives you a good start, and reduces the number of booklets you would have to write away for.

2 (a) From your own and your friend's knowledge, or from travel brochures, make a list of all the towns in Britain which are holiday resorts.
(b) From an atlas, trace an outline map of Britain.
(c) On your map, show as dots, without naming them, the holiday resort towns.
(d) From an atlas, draw on to your map the 15°C summer isotherm. Shade lightly, in different colours, the area warmer than 15°C in summer, and the area cooler.
(e) Work out the percentage of holiday resorts which have average summer temperatures over 15°C.
(f) Now, from an atlas again, draw on your map the 500 mm summer isohyet.
(g) Shade, in stripes, the area with over 500 mm of rain in summer.
(h) Work out the percentage of holiday resorts which have less than 500 mm of rain in summer.
(i) Work out the percentage of places which have *both* over 15°C summer temperatures and less than 500 mm of rain in summer.
(j) Do the same exercise for places which have both summer temperatures of less than 15°C and more than 500 mm rain in summer.

3 Carry out a survey of a park near your school. The following investigations might be possible.

(a) How big is it?
(b) Roughly how many people per week use the park?
(c) Make a list of the facilities provided.
(d) Investigate the opinions of people, including children, in the park. Questions you might ask are:
 (i) How many times have you visited the park in the past week?
 (ii) Did you just come to relax quietly?
 (iii) If not, which facilities did you use?
 (iv) Do you think more facilities are needed?
 (v) If so, which?
(e) Investigate the opinions of people

in the neighbourhood. Questions you might ask are:

(i) Do you use . . . . . Park?

(ii) If so, how often, daily? weekly? occasionally?

(iii) Do you think another park is needed in this area?

(iv) If so, where would you put it?

(f) When you have carried out your survey, draw up a group plan for leisure in your area. Use maps, photographs, drawings, graphs as well as written statements.

(g) Find out if there is a development plan for your area. If there is, compare its plans for leisure with your own.

4 Organise a class debate, with some speaking in favour, and some against, the motion that, 'All mining and industrial activity in National Parks should be prohibited'.

6 And so make a city here

Finding a village site

Figure 82 shows a picturesque but ordinary English village. There is nothing very unusual about it. Yet there has been a village here for over a thousand years! Most of the villages in England were started sometime between AD 400 and AD 900 by the Anglo-Saxon or Norse invaders. The Norsemen also founded many of the villages of Scotland.

In England, the first Anglo-Saxons settled mostly on the south and east coasts because they were nearest to where they came from. Gradually, they spread westwards until their villages covered most of what is now eastern England.

Imagine you are an Anglo-Saxon chieftain (or chieftain's wife – she probably has a fair say in what follows). You have decided to leave your father's village because there are now too many people to live off the farming land around it. You probably also think your father's ideas are old-fashioned and you could do better if you were in charge. So you set off, together with your family and those others who wish to come with you. You head west, because the land to the east has already been taken up.

As you leave the cultivated land, you enter thick woods. Most of England was forested at this time. You know these woods a bit, as you and your people have collected firewood or building wood from them, and perhaps hunted in them. Even so, they are rather frightening, and the further you go from home the more frightening they seem.

Now you are looking for somewhere to settle. You do not wish to be too far from your father's village. This is partly because of your fear of the unknown, and partly so that you can visit your family and friends fairly frequently. You will need to find a place where there is a nearby spring or stream for a regular water supply. You will need good arable land to grow cereals. You will need some pasture for your animals and a supply of building materials and fuel. Although you need a water supply, you do not wish to build your village in a marshy area or one which might flood.

Figure 82 Finchingfield, Essex

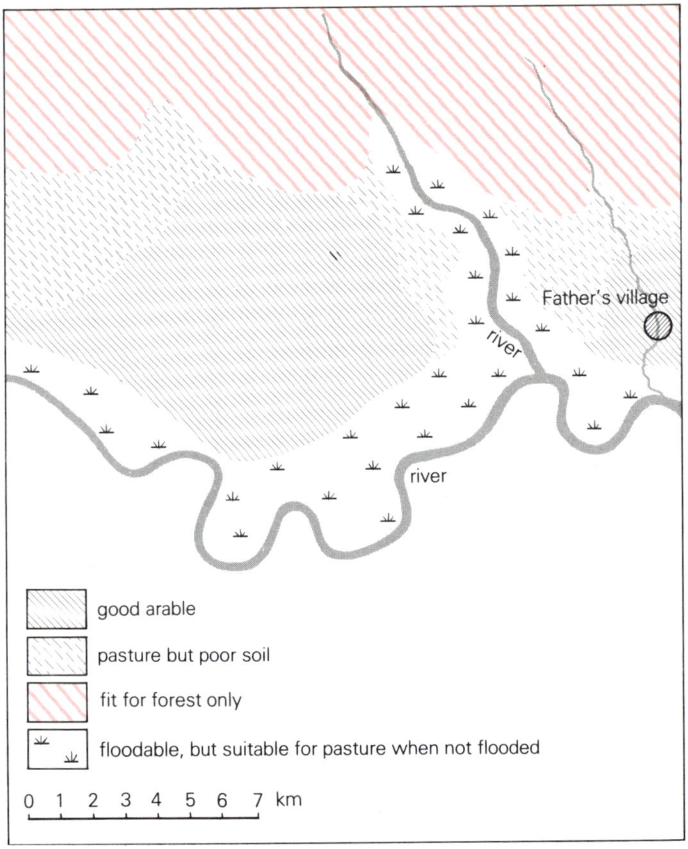

Figure 83 (above) *All kinds of monsters were very real to the Anglo-Saxons*
Figure 84 (above, right) *Possible village sites*

Now look at figure 84 and decide which place you would choose for your village. In making your decision, you are better off than a real Anglo-Saxon. You have a map and you can cover the whole area looking for the best **site**. A site is a piece of land on which to build . . . . Even though the Anglo-Saxons may have made scouting expeditions before actually setting out, their knowledge of the countryside would be very sketchy. When they had travelled until they were tired, they would perhaps choose the best site *they had found*. This may not have been the *best in the area*.

There would also probably be disagreement about which *was* the best, even of those they had seen. You will know from chapter 1 that people see things differently.

The place you choose will depend on how important you decide the eight factors are. To remind you, here is a list of them:

nearness to: father's village, water supply, arable land, pasture land, building material, fuel; and

absence of: marsh or floodable land: werewolves, hobgoblins or other forest 'unknowns'.

For example, you may think that nearness to water supply is more important than nearness to your father's village.

If you write down the list of factors in your order of importance, and then write down, against each of them, the distance from your chosen site (using the scale on the map), you should then be able to give the reasons for your choice fairly clearly.

Luck comes into it too. The real chieftain may have had the luck to find a good site, or he may have been unlucky and missed the best site in the area. We cannot do the exploration in a book, but we can imitate, or simulate it. We can introduce luck by throwing dice, so, when you think you have chosen a site, throw a number, then look at figure 85 to see whether you are lucky or not. If you are lucky, you can stay on the site chosen. If you are unlucky, you must move on and find the next site.

Figure 85 *Chance table*

Distance from father's village	lucky	unlucky
Under 7 km	1,2,3,4,5	6
7 to 14 km	1,2,3	4,5,6
Over 14 km	1	2,3,4,5,6

Parishes

In this simulation, you have found out something about how Anglo-

Saxon settlement spread across the country, and what were the influences on **locational decisions**.

There were no rules in the game against your settling too near your father's village. This was to avoid it becoming too complicated. Yet, really, there should have been such a rule.

In those days, there was not much trade and people lived on what they grew. Obviously, it would take a certain amount of land to feed a certain number of people. If another village was too close, there would not be enough land to go round. The land around the village, on which it depended, eventually became the **parish**. The parishes we have today have survived without much change for hundreds of years.

1 If the land were absolutely flat, and of the same farming quality all over, what shape do you think the parish would be? Take it that all the farmers live in the village.

2 If there were several parishes, what sort of arrangement would be needed for them to be as close together as possible? Take it that they are all the same size, that the land is flat and all of the same quality. Try using coins of the same size to represent the parishes. Shuffle the coins around until the spaces between them are as small as possible.

3 When you have answered question 2, you can try to think what would be the pattern formed by the villages. The villages would be dots in the middle of the coins.

Figure 86 shows what parish boundaries should be like if the land were the same all over. As you know, land is not the same everywhere. In deciding where to site your village at the beginning of this chapter, you were looking at a very varied landscape.

You have to think of the pattern in figure 86 as an ideal, or **model**, one. If you look at actual parish shapes,

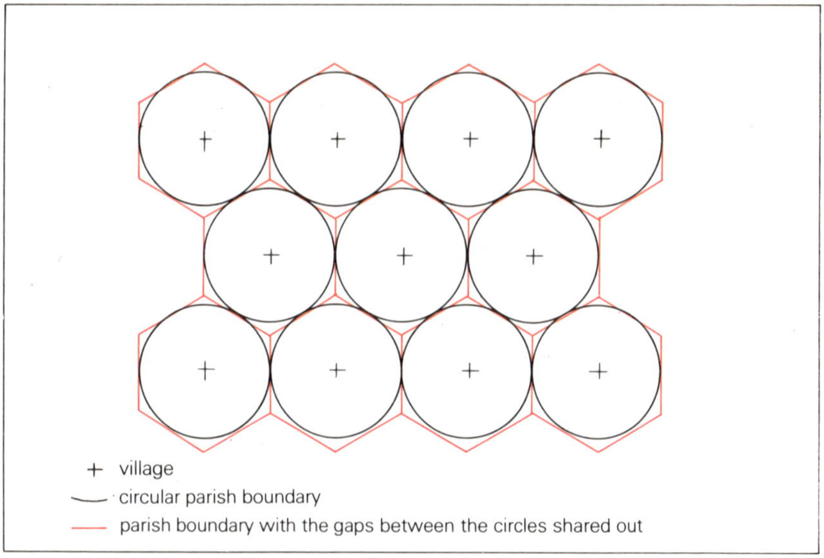

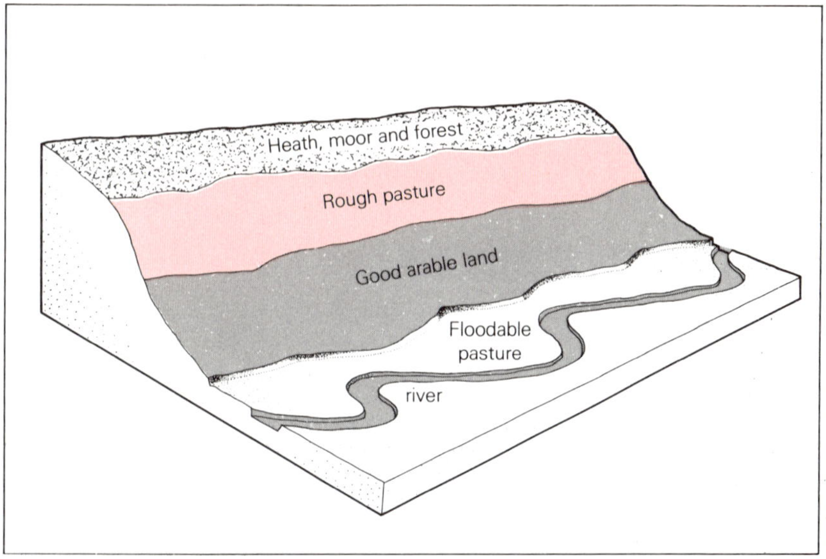

Figure 86 (above)
Figure 87 (below) Possible land use on a scarp slope in Anglo-Saxon times

you could start by saying, 'all parish shapes should be hexagons, so why are these different?' That is to say, you could use the ideal, but unreal, pattern as a starting point when looking for an explanation. Using a model, or **theoretical**, approach to finding explanations is one that is used a good deal today in solving many kinds of problems. We shall meet it again later.

4 If there are seven parishes in the area shown on figure 87, what shape would each be in order to give every parish an equal share of each type of land?

5 Look at figure 88. Is there any similarity between the parish boundary shapes and those in your answer to question 4?

6 If the shapes are not as you expected, can you think of any reasons why?

In looking at these actual parish shapes, you are again using a theoretical approach. In this case you are saying that if you are given the information about land use and the

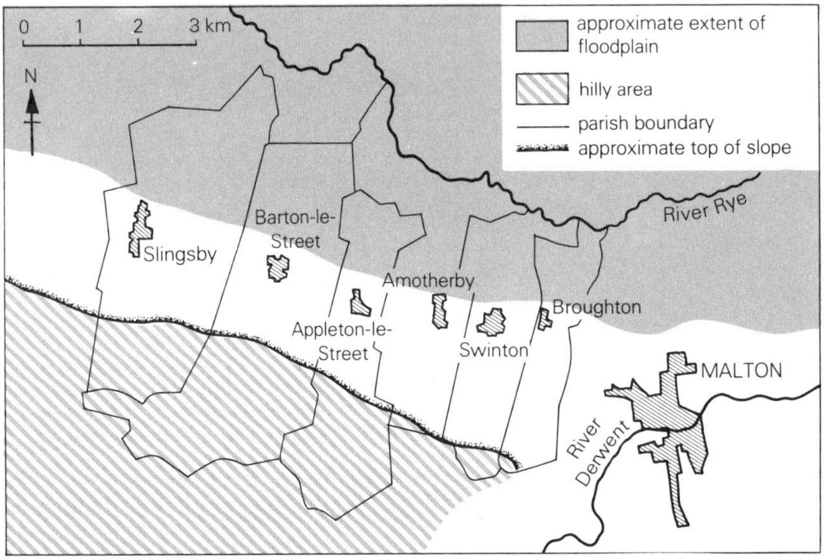

Figure 88 (above) Parish boundaries at the edge of the Vale of Pickering
Figure 89 (below)

number of parishes, the parish ought to be a certain shape. If the shapes are actually like that, you have probably explained them by your **model**. If the shapes are not as you thought they should be, you are in a better position to start looking for explanations than if you were starting from scratch.

7 Look again at figure 87. Where do you think the villages would be inside each parish?

Types of rural settlement

By now you may have noticed that rural settlement does not consist entirely of villages. In figure 88 there are small groups of houses, with names like villages, but smaller. These are **hamlets**. You will also see some isolated houses, which together are known as **dispersed** settlement.

8 Make a tracing of the **rural settlement** in a ten-kilometre square on any one-inch or 1:50000 map in England. Mark each village with a circle, each hamlet with a triangle and each isolated house with a dot. If there are only two or three villages in each ten km square, you could say that the settlement is mostly dispersed.

In question 8, you will have had difficulty in deciding the difference between a village and a hamlet. A village provides services, like shops, schools and churches, for the surrounding area. A hamlet is just a collection of farms, with perhaps a small shop or garage. The only way to find out for sure which is which, is to go and see. For question 8, you have just had to guess from the size on the map. Even if there are more than two or three villages on your tracing, there are likely to be farms in between them. In many parts of Britain these are fairly new. They were built when the owning of land in common by the villagers was ended, and fences, particularly hedges, were put up. After **enclosure**, each farmer owned a particular area of land, and set up his house in the middle of it. Before that, the farmers had lived in the villages and had farmed strips of land scattered all round in the open fields.

Although much dispersed settlement has grown in the way just described, there are many areas where the settlement is mostly dispersed and where there are few villages. One of the reasons for this can be found in figure 89.

9 In figure 89 how would you describe the landscape?
10 What is the crop that the farmer is dealing with?
11 How many fields can you see which would be suitable for growing crops?
12 In fact, how many real *fields* can you see?

In the highland zone of Britain, there are areas of infertile land which cannot produce enough food to support a whole village, so there are isolated farms making a living as best they can. This does not mean that all the land in Highland Britain is infertile, or that all settlement is

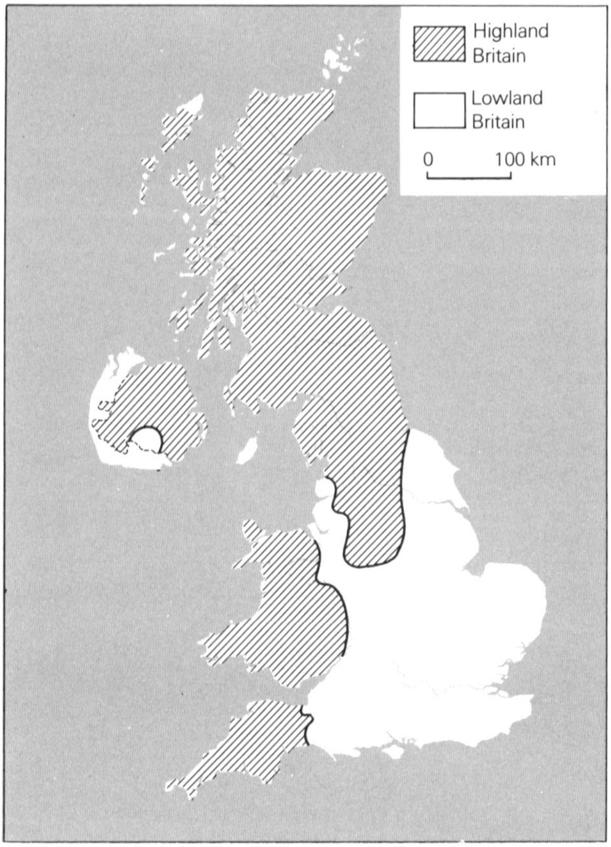

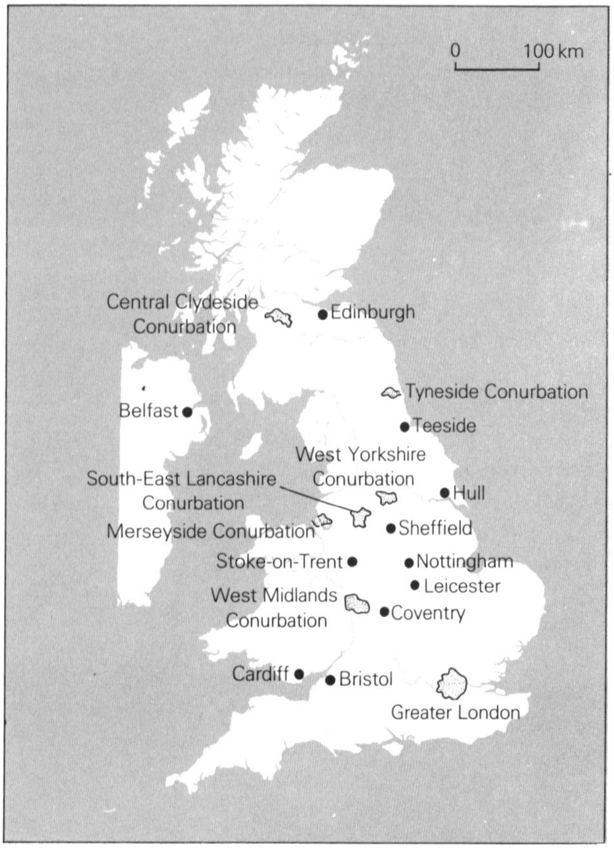

Figure 90 The highland and lowland zones of Britain

dispersed. It is just that this is one of the reasons why villages have not developed in some areas.

Another example of dispersed settlement can be found in the Fenland (pages 11 and 12). The draining of the fens on a large scale was not begun until the seventeenth century. Since then, large areas have gradually been made useful for agriculture as methods of draining them have been invented. As the drainage proceeded, so did the building of farms. As we have seen, the villages in the area had been started hundreds of years before, but in the new lands there were very few of these.

We have spent some time studying rural settlement. One of the reasons for this is that there are more rural settlements than any other kind in the UK.

Orders of settlement

It may seem obvious to say that there is a small number of large places, a larger number of middle-sized ones and a larger number still of small settlements. Obvious, yes, but it is another example of some kind of order in the way people behave.

Look at the list of the largest settlements in the UK shown in figure 92. We can only show the largest because there are far too many settlements altogether.

13 Do you remember what a conurbation is? If not, turn back to page 18 to see.

You can see that London stands alone at the top. Then, after a very big gap in size, come two conurbations together, then not quite such a big gap, and so on. After the conurbations, the sizes of the main cities are much closer together. Figure 91 shows the places named in the list.

Figure 91 UK: largest settlements

Figure 92 UK: population of largest settlements

	Population
Greater London	7452000
South East Lancashire conurbation	2393000
West Midlands conurbation	2372000
Central Clydeside conurbation	1728000
West Yorkshire conurbation	1728000
Merseyside conurbation	1267000
Tyneside conurbation	805000
Sheffield	520000
Edinburgh	454000
Bristol	427000
Teesside	396000
Belfast	362000
Coventry	335000
Nottingham	301000
Hull	286000
Leicester	284000
Cardiff	279000
Stoke-on-Trent	265000

14 If you draw a graph of the settlement size, it may help to make the idea of order in settlement clearer. (See also chapter 5 in *Spotlight on Europe*.)

First, put the places in **rank order**. This has already been done in figure 92, but it means calling the largest place no. 1; the second largest no. 2; and so on. Draw the graph with the size of the settlement, that is, the size of its population, up one side and the rank along the bottom. Figure 93(a) shows you how.

Mark each place with a dot, then connect the dots. The result should be roughly the shape shown in figure 93(b).

Now we have seen that there are not so many urban settlements as rural, so what is their importance?

Well, you live in a town don't you?

Urbanisation

Some readers will be able to say 'no' if asked whether they live in a town, but eight times out of ten, the answer will be 'yes'. This is because almost eight out of ten people in the UK live in urban areas. A way of showing divisions is shown in figure 94. In this diagram, 3·5 degrees on a protractor equals about one per cent. The circle is divided using this scale. For every one per cent, multiply by 3·5, and that is the number of degrees you use to show each segment of the circle.

15 Using a protractor, find out roughly what percentage is shown by each segment of figure 94.

16 From figure 95 draw divided circles for each of the countries of the UK.

17 In which country would the statement, 'You live in a town' most likely be true? In which country would it be least likely?

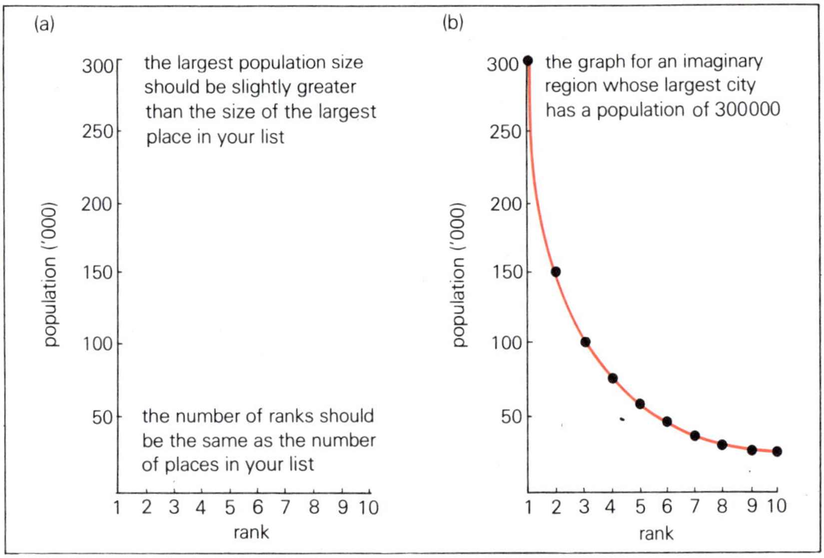

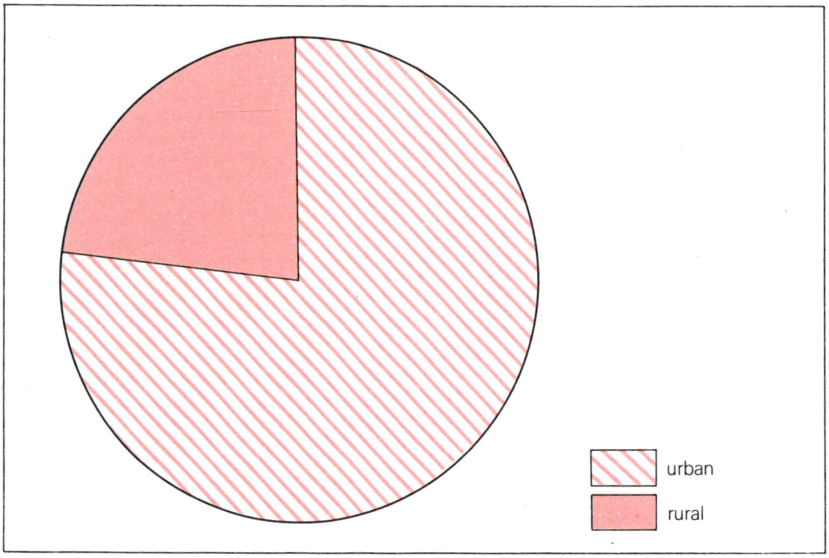

Figure 93 (above) (a) Axes of the rank/size graph (b) Rank/size graph
Figure 94 (below) UK: proportion of urban and rural population

	Urban %	Rural %
England and Wales	78	22
Scotland	71	29
Northern Ireland	55	45

Figure 95 Northern Ireland, Scotland, England and Wales: rural/urban percentages

Provision of services

Study figure 96 and you will realise quickly, from your reading of chapter 4, that it shows one kind of city region. In this case it is a workplace map.

You know from chapter 4, that although eighty per cent of people in the UK live in towns, a still higher percentage uses them. In the past, the main difference between a town and a village was that most of the people in towns did not work in agriculture. Instead, they provided services, like schools, shops, hospitals, or they worked in industry.

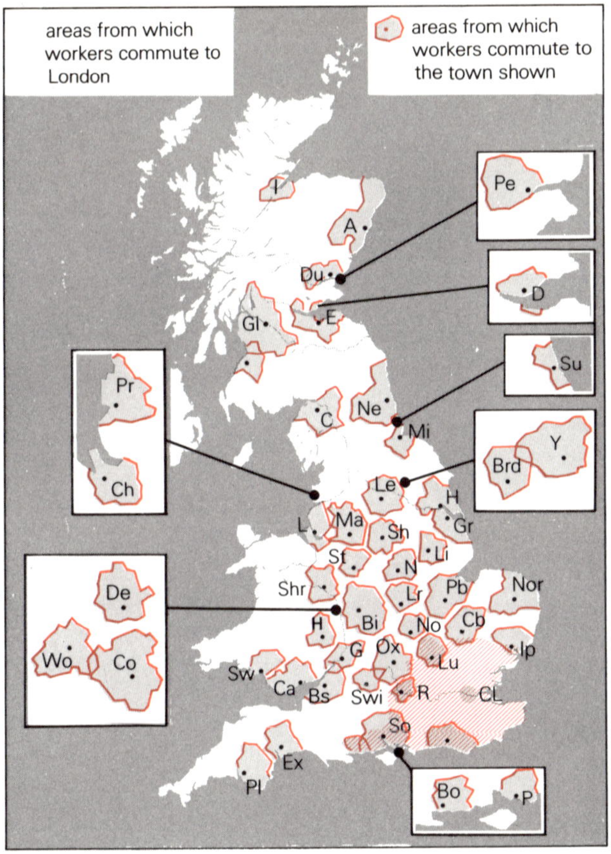

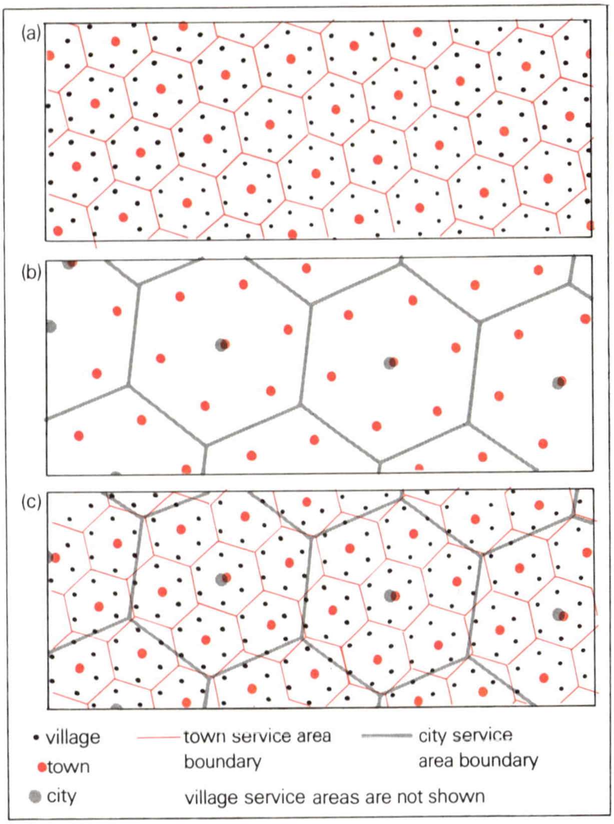

Figure 96
Figure 97 (right)

Nowadays, many people who live in villages work in towns and **commute** to work. There are very few people now working in agriculture. Even the people who work in the villages mostly provide services for the commuters. You might almost say that the villages are extensions of the towns, particularly since they enjoy all the conveniences of towns, like electric light, fast transport, mains water and sewage. This is why the idea of a city region is such an important one.

Many of the people who use the shops, offices, recreation places and other **services** of a town will have to come from outside it. The bigger the town, the more customers will be needed. That is to say, to keep a town of a particular size going, a city region of a certain size will be needed.

The pattern formed by the towns and the shape of their **service areas** works in the same way as the Anglo-Saxon villages. Do you remember what these shapes were and how they came about? If not look back to page 67 and figure 86. The patterns formed by towns and their service areas will be similar if we say that the land is exactly the same all over.

In chapter 4 we saw that city regions 'nest' inside each other. The same is true of service areas, which are almost the same thing as city regions. Small towns have service areas, but they themselves are part of the service areas of larger towns. These in turn are inside the area of a city and the people of a city will occasionally need the services of a **regional metropolis**, the capital of region. Finally, everyone in the country at some time or other needs the services of the capital city, though this does not necessarily mean they have to go there for them.

Using the model pattern of settlements and shape of service areas, one way in which nesting could occur is shown in figure 97. Many more arrangements are possible and you may come across them in another book.

At this point three things may be puzzling you:

(a) What is the difference between a town and a city?

(b) What is the difference between a city region and a service area?

(c) The UK land is not the same all over, so what affects the location of towns besides the **theoretical** pattern?

For practical purposes a city is simply a bigger town, with a larger service area. Similarly, a metropolis serves a wider area than an ordinary city.

There is very little difference between service area and city region. Service areas were introduced to you because almost all the towns

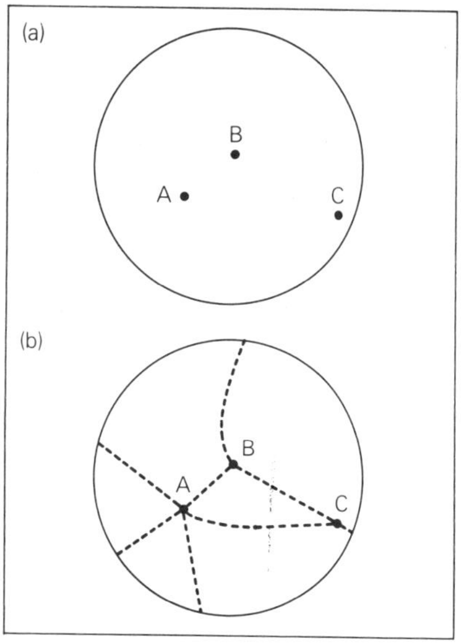

Figure 98

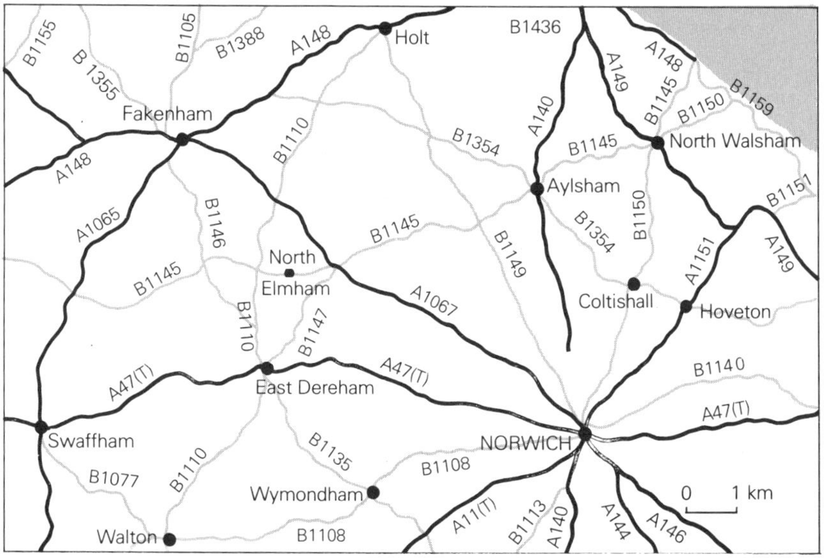

Figure 99

in the UK were once service centres, even if now they are industrial centres as well. Before the Industrial Revolution, towns were places where people went to buy and sell. That is what is meant by 'service centre'. Today, *all* towns serve this purpose, even if they are better known as industrial towns or holiday resorts. The service centre idea leads us to answer the third question. Since nearly all towns developed for this purpose, the main thing needed for a good location was the ease with which people could get to them from outside.

Accessibility

18 In figure 98 imagine that the circle is an island or circular region. It is completely flat, and the people are evenly spread all over it, except in the towns. The three dots, A, B, and C are towns.

(a) Map (a) shows the region with no roads. Everybody has to walk everywhere. Which town is easiest to get to for all the people? Remember, it has to be *all* the people. Obviously some places are easier to get to for the people nearest them.

(b) It was becoming too difficult carrying goods to and from market, so now roads have been built (map (b)). These make it much easier to carry heavy loads. Which town is now the easiest to get to for *all* the people.

The ease with which people can get to a place is called its accessibility, and the *most accessible place* in an area is *the one to which most people can travel most easily.*

You will have guessed from question 18 that the meeting of roads often creates greater accessibility. You can get a rough idea, therefore, of which are the most accessible places by counting the roads that enter a place. For instance, in figure 98 (b), place A would have a score of five; place B, two (a road that passes through is the same as two roads entering it); place C, three. The trouble with this method is that it counts all roads as equal. Surely a main road can carry more traffic over greater distances than other roads. Perhaps a better measure is to give points say three for a main road; two for a B road and one for other roads. You could give two points for a railway station.

19 Using the method just described, find the most accessible place, apart from Norwich, in figure 99.

One difficulty you will have is deciding whether roads 'enter' a place, or meet just outside it. For example, to the north west of Fakenham, the B1355 meets the A148. The meeting is outside the town, so strictly we should only count three for the main road entering. But surely the B road also aids Fakenham's accessibility?

There is no answer to this difficulty, you will just have to use your judgement. The main thing is to use the same rules, whichever you invent, for every place.

What causes roads to meet at a particular place? If the land is flat, the population evenly spread over it and there are no obstacles to road building, we are likely to find the roads meeting at the centre of hexagonal service areas. We shall be back to our theoretical pattern! The nearest we have to this in the UK is probably East Anglia.

20 Place a piece of tracing paper over figure 99 and mark with a dot the centres of the following places: Holt; Fakenham; North Wal-

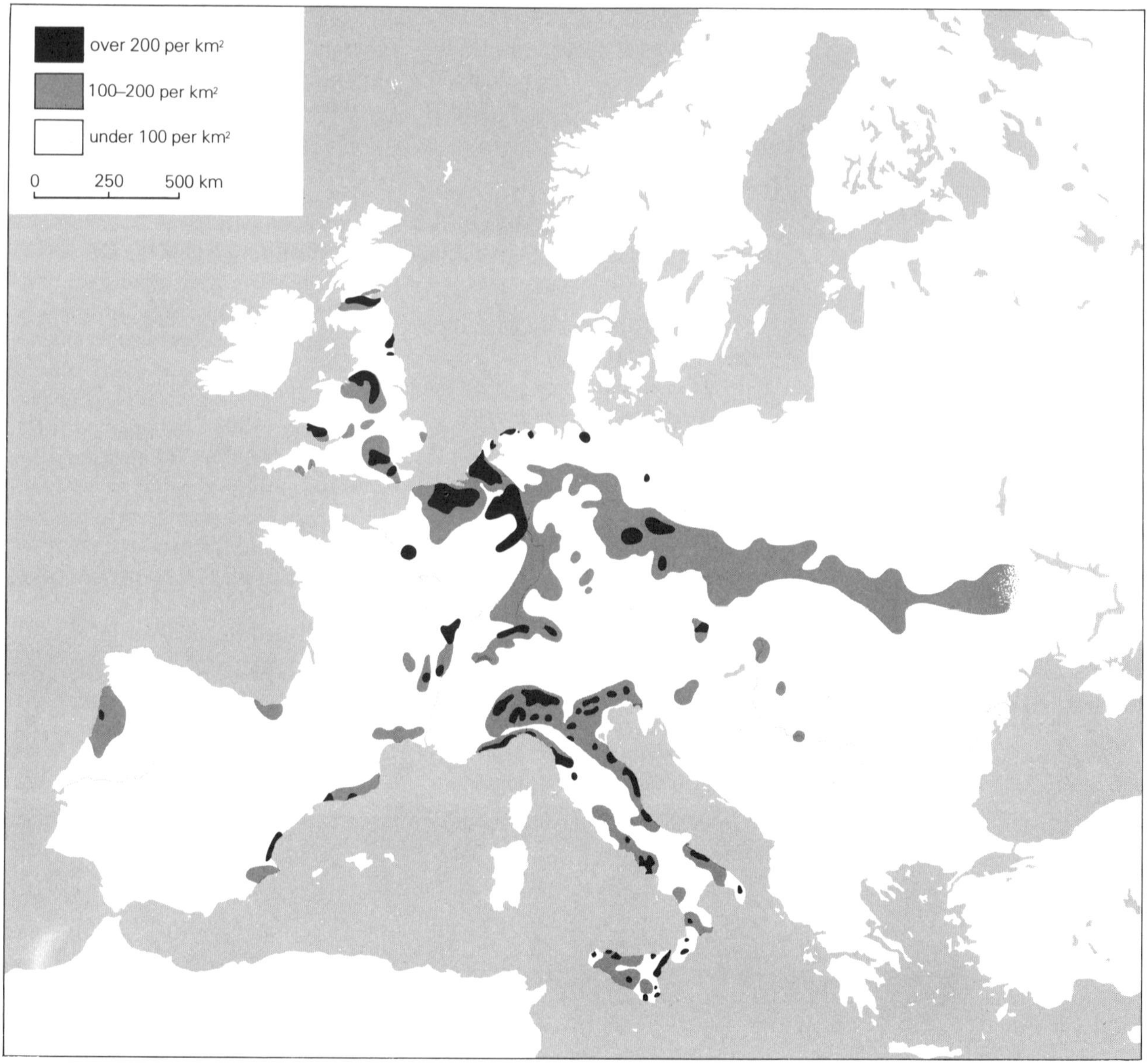

Figure 100 Europe: population distribution

sham; Aylsham; East Dereham; Hoveton; Swaffham; Wymondham; Norwich.

Is the pattern formed anything like the theoretical one? Even if you do not think it is, the towns are at least fairly evenly spaced.

If the roads are obliged to go round hills, or through gaps in the hills, to avoid steep slopes; or if they meet to cross a river; or if they follow any kind of easy path; or avoid any obstacles, then the ideal pattern will be changed.

It is most important to remember that these places did not grow *because* of the gap, or the routeway formed by the valleys, or the bridging point. They grew because they were accessible. This is because people wanted to travel to them. The relief which influences roads does not create them. There will only be roads if there are people to use them.

21 If there are two gaps through a range of hills within less than a kilometre of each other, are they both likely to contain towns? Give reasons for your answer.

One could ask a similar question about a river which has several places suitable for building bridges.

Apart from relief, there are other things which might alter the theoretical pattern formed by towns. For example the presence of coal, or occasionally, other minerals. During the Industrial Revolution the population of the coalfields became so great that the towns had to be packed closer together than in other areas. But even on the coalfields, it was usually the most accessible towns that grew into the biggest cities, and the most accessible villages that grew into towns.

Conurbations again

What has affected the location of the conurbations? To some extent, the coalfields have. But there are two big conurbations which are not on coalfields. From chapters 3 and 4, you will realise that these two are where there are most people in Britain. We have seen that this central area of Britain is where there are the most jobs, the biggest market, and so on. Another way of putting it would be to say that it is the most accessible part of the UK.

The 'most accessible' place is the one that is easiest to get to for most of the people. We have already seen that most of the people of the UK can most easily get to the London and West Midlands conurbations. Now look back to figure 100.

Figure 101 (below, left)
Figure 102 (below, right) British Isles: main ports and harbours

22 Which UK conurbation is nearest to most of the people of Europe?

Ports

Before leaving the question of location of settlement, there is one more special kind to look at.

23 In figure 101 which of the good harbours do you think is most likely to become an important port? Give reasons for your answer.

In the British Isles, the best harbours are in thinly populated areas, such as the Highlands of Scotland, the south-west peninsula of England, west Wales and south-west Ireland. Figure 102 shows the main ports of the UK plus the main areas of good natural harbours.

In the heavily populated part of Britain, only Southampton has a very good natural harbour. Milford Haven is the only natural harbour away from the main centres of population which has a really big tonnage of ships entering and leaving. Milford Haven is a very special case. Even with the large oil trade, there is no large town which could be called an important port.

The main ports of the UK have changed, however. If you have read any stories of piracy on the Spanish Main, you will perhaps remember that in those days, the many good natural harbours of the south-west peninsula were also important ports. The harbours are still there, and many of the towns, but they are nothing like as important as they were. In Elizabethan days, there were no coalfields and no conurbations. The population was distributed very differently.

PROJECTS

1 Look at a copy of OS 1:50000 sheet 130 (Grantham) In the area

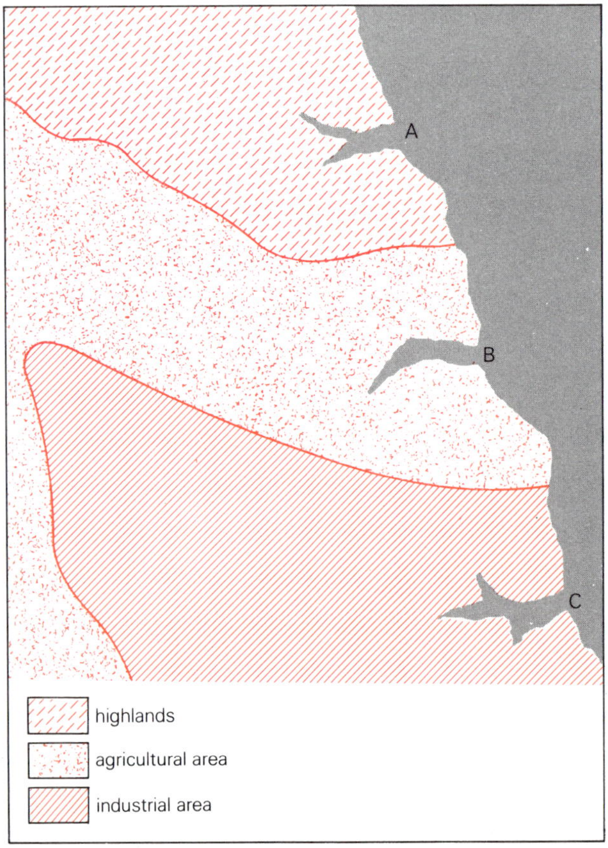

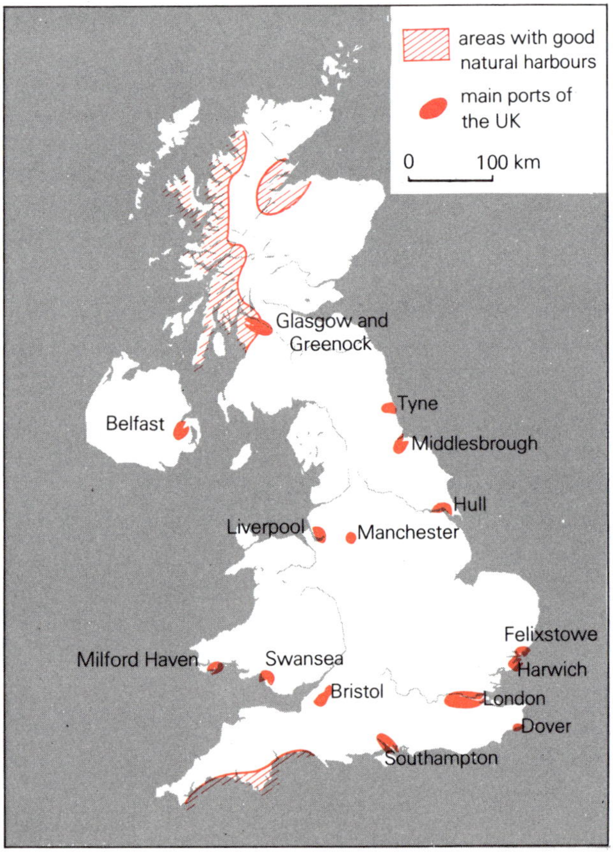

between eastings 07 and 20, and northings 25 and 35:

(a) How many contour lines can you see to the west of the course of the old railway?
(b) How many can you see to the east of it?
(c) Find the village which is furthest away from a contour line. How far is it, roughly, to the contour line?
(d) How would you describe the settlement in that part of the map which has no contour lines?
(e) Which area of Britain is this map extract from? You will have to guess the answer from the names to the east of most of the villages.
(f) Can you remember why most of the settlement in the flat area is dispersed? If not, turn back to page 69 for the answer.

2 Investigate the origin of your nearest town, or city. You can start this investigation by mapping the physical geography of the town centre area.

First, use a 1:25000 map. Trace the outline and then put a cross where you think the centre of the settlement is. Next carefully trace the contour lines in and around the village, then trace any watercourses or springs.

How far is the centre of your town from the nearest water? Is it built on a slope? If it is very near to a water supply, is it likely to be flooded? Sometimes, towns are sited on very low rises in the ground, often on **river terraces**. These frequently do not show up on contour maps because they are not high enough.

Because accessibility is important for towns, you can go on to map those items which you think might have affected accessibility. For example, you can map roads, gaps in hills, and so on. Towns are easier to get to for most people. That is why they are towns. This means that you should not find it too difficult to visit a town, even if you are not one of the eighty per cent who live in one.

In the town, examine the site. This may reveal things which the map does not show, such as small relief features like river terraces or low cliffs, which may have helped defence, or kept the towns free of floods. Remember, you are looking for the site as it was when the town first began to grow. This may not be in the same place as the present town centre.

You can check on the roads which you mapped. Are they the original roads, or were they built later?

A visit to the museum may give you some extra information about the origin and growth of the town. As with villages, the study of the origins of towns could be a book in itself. The ideas here are just to get you started.

3 Conduct an accessibility survey of your town, or your nearest town. This could take two main forms, first an historical survey; secondly a present-day survey.

As indicated in the previous project, the historical survey could investigate the communications during the past. Visits to the museum or local library; looking at old maps and historical documents; talking to local historians and so on will give you an idea of how the accessibility has changed over the centuries.

The present-day survey is rather more difficult if you live in a city. Perhaps you should confine it either to the city centre, or, if this is too big, to one of the larger suburban centres. Some of the things you can do are as follows:

(a) Count the number of people emerging from the railway station at rush-hour. If it is a big station, this will need several people to count and, preferably, hand tally counters to help.
(b) Count the traffic on all the roads into the centre. If the centre is not too big, a whole class might be able to throw a cordon round it and do the whole thing at once. If this is not possible, it is still fairly accurate to do each road separately, so long as you do them for a fairly long period (say half an hour) at the same time of day.

You now know certain things about your town's accessibility, but, so what? The best use you can make of your information is to 'weight', the various roads and railways more accurately than in question 19. Once you have counted traffic and railway passengers, you may decide that a score of three is not enough (or too much) for a main road, and that two is not the right score for a railway.

Using the result of your counts, calculate the best score for each line of communication. Then do question 19 again, using your own **weighting**. You may not have encountered this word used like this before. Its meaning is fairly clear from the way you have used it in this exercise and in question 19.

Once you have a weighting system that you think is accurate, you can use it to compare the accessibility of your town with others which you can only 'visit' on a map.

A word of caution here. Your town may be unusual, so if you get the opportunity to do counts anywhere else, work out the weightings for the different types of communication. See if they are the same as for your own town.

7 What a place to live!

Figure 103 (above)
Figure 104 (below) This picture was taken in Belfast in August 1971

Regions within the city

You might be in trouble with these two characters if you were found in the wrong part of Londonderry or Belfast. What is meant by the 'wrong part' of the city? Figure 105 shows it very clearly and very tragically.

Of course, not all the deaths mentioned were of people found in the other side's territory. Many were gunmen fighting inside their own areas. The important thing is that the regions exist. They are known by everybody in Belfast and are clearly shown in figure 105. They are now only too well-known outside Northern Ireland. Shankill, Falls and the rest are household names to millions.

1 Andersonstown; Crumlin Road; Shankill Road; Falls Road; Divis Flats; Ballymurphy; Ardoyne. These are a few of the better-known names in Belfast. Look at figure 105 and find them. See which are Catholic and which are Protestant areas.

What if you should actually live in one of the other side's areas? In untroubled times a few people of the 'wrong' religion have managed to exist peacefully with their neighbours. But when the bitterness overflows, extremists on both sides may force out innocent families at gunpoint and burn their homes behind them (figure 104).

Once this sort of terrorism occurs, fear takes hold and people leave their homes. By this time hatred may have grown so much that they burn their own home as they leave, to prevent the others using it after they have gone.

Even in peaceful times there has always been very strong religious segregation in parts of Belfast. In Shankill, 99 per cent of the residents were Protestants before the disturbances started; in nearby Clonard, 98 per cent were Catholic. The boundaries between the areas were very sharp, often only the width of a road.

2 Look at the scale of figure 106. You will see how narrow the dividing line is between the two communities.

The word 'communities' is used in question 2. The existence of these very strongly segregated areas is one example of a sense of community. In Belfast, this sense is exaggerated, which is why it was chosen as an example. In other places people have a sense of community, but they do not usually fight because of it.

3 Do you know of any communities around your own home?

Of course, the disturbances in Northern Ireland were not caused by its

Figure 105 (opposite) A map from the Sunday Times, *July 1972*

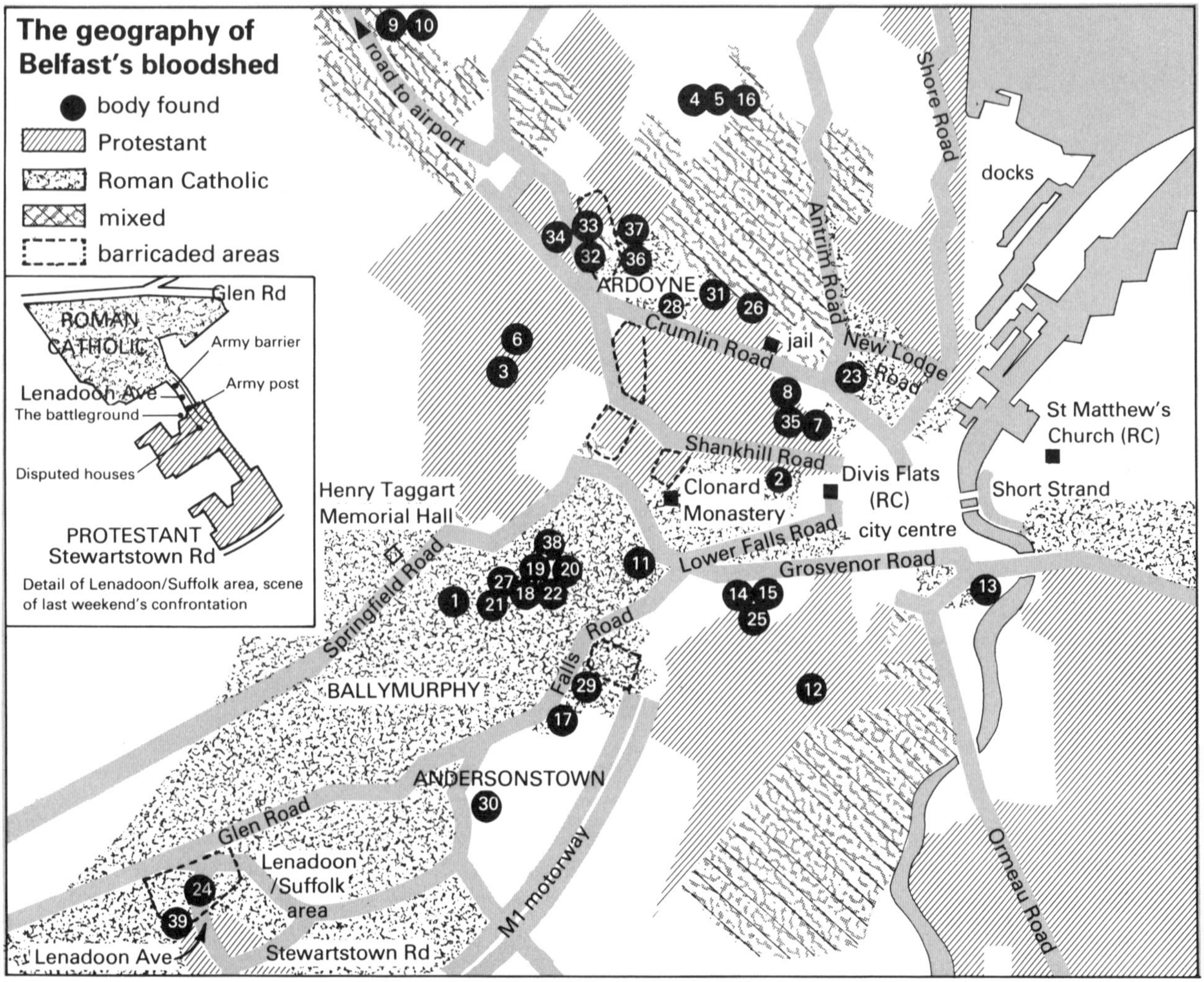

The deaths the ceasefire didn't stop

In just 19 days, about 40 people have been killed in Ulster—and 11 of those days were included in what was regarded as a ceasefire. The list below is of the victims known up to yesterday morning. Most were killed in Belfast. Numbers refer to locations on the map; P denotes Protestant, C Catholic where religion is known.

JUNE 27 – Truce begins

1: Bernard Hornsey (C). Father of five, shot in his car.

JULY 1

2: Daniel Hayes, 43 (C). Dumped from car.

3: Paul Jobling, 19, from Co. Durham. Found hooded and gagged on a rubbish tip.

JULY 2

4: Hugh Clawson, 30 (P) Lorry driver.

5: David Kisher, 30 (P), bus driver. Shot through head.

6: Gerald McCrea, 27 (C), of Andersonstown. Body thrown from car.

7: James Howell, 31 (C). Hooded and shot in McCrea's Mercedes.

JULY 3

8: John Patrick O'Hanlon, 35 (C). Hooded and shot.

JULY 5

9: Malcolm Orr, 20 (P) and

10: Peter Orr, 19 (P), brothers.

JULY 7

11: Samuel Robinson, 19 (P). Shot in head in car.

JULY 8

12: Lawrence McKenna, 22. Shot from passing car.

JULY 9

13: Gerald David Turkington, 32. Shot.

14: Joseph Francis Fleming, 29 (C).

15: Brian Patrick Macmillan, 21 (P). Shot in car.

16: David Andrews, 31 (P). Shot.

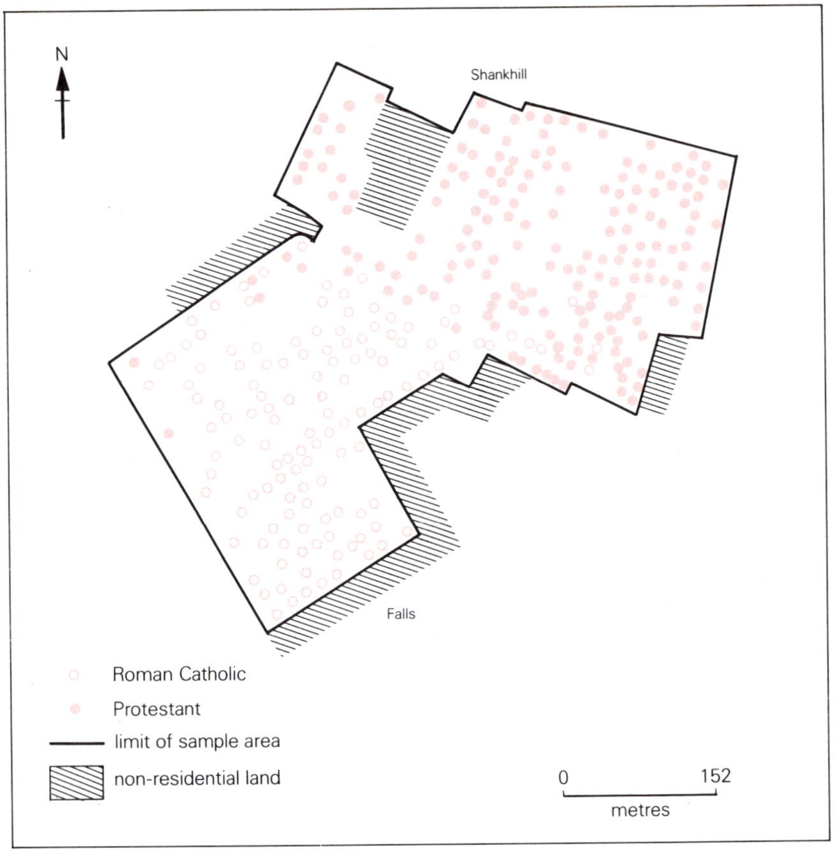

Figure 106 Parts of the Shankill and Falls areas of Belfast

geography. Historically, there has been a long conflict between the two sides. Economically, unemployment is high. There are many people who earn little enough to make them dissatisfied and who are likely to join the extremists. It is no accident that the areas with strongest religious segregation are working-class areas.

Social regions

How do you know whether an area of a town is 'working-class' or not? Or, for that matter, any other class? The best way is to find out a great deal of information and then use a computer to work out what it means. Figure 107(a) was drawn in this way, but as you can see the map itself is very easy to understand.

4 Give the names of two high-status and two low-status areas in Belfast.

5 What do you think is meant by the word 'status'?

6 Draw two columns, one headed 'high status' and one headed 'low status'. Put the following items into one column or the other. All these items were used in drawing figure 107(a): large number of persons per room; large number of semi-skilled and unskilled manual workers; large number travelling to work by car; high percentage of people with driving licences; high percentage of people out of work; high percentage of people aged 16 to 19 in full-time education; large number of people journey to work on foot; large numbers of managerial and professional workers.

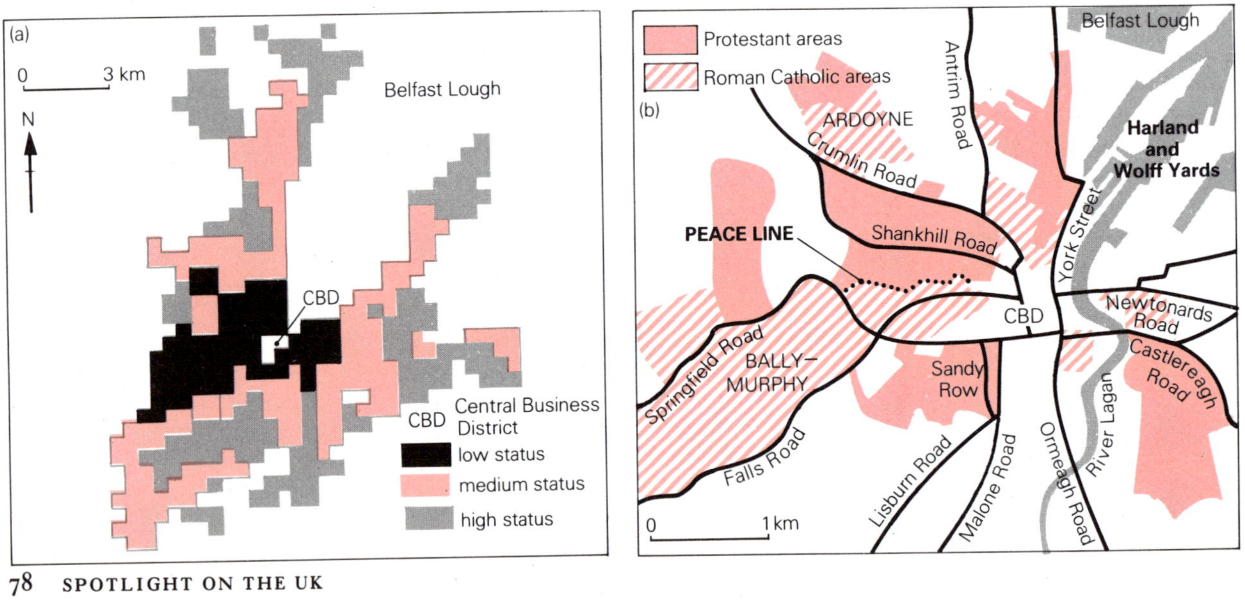

Figure 107 (a) Social areas of Belfast
(b) Main roads and districts of Belfast

Another, less accurate, way of discovering social class is to look at the types of housing.

7 Look at the two photographs in figure 108. One is in a 'high-status' area and one in a 'low-status' area. Which do you think is which?

Another **indicator** of social status is the **density** of population (see chapter 1, *Spotlight on Europe* in this series). This means the number of people who live on a particular area of ground. For example, if 100 people live in an area of one hectare, then the density is 100 per hectare. If 100 people live in an area of two hectares, the density is 50 per hectare.

8 What is the density of population if 300 people live in an area of six hectares?

9 Do you think a high density of population is likely to show a high-

Figure 108

Population 1971		
Wards	Area (hectares)	Persons
Belfast County Borough	**6400·16**	**362 082**
Clifton	627·27	43 914
Court	43·30	4769
Cromac	461·75	19 392
Dock	74·06	9650
Duncairn	611·08	30 998
Falls	326·58	33 681
Ormeau	511·53	36 373
Pottinger	560·09	38 537
St Anne's	473·08	20 018
St George's	54·23	8860
Shankill	555·64	32 850
Smithfield	37·64	4646
Victoria	1209·61	31 184
Windsor	421·28	20 945
Woodvale	433·02	26 265

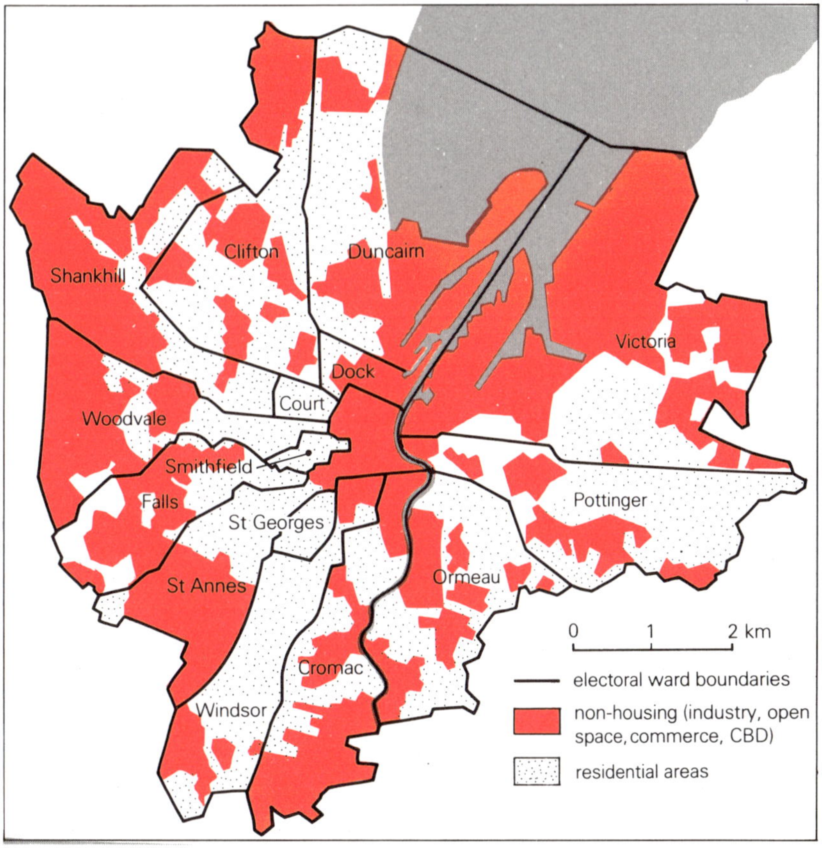

Figure 109 Belfast: electoral wards, residential areas and population

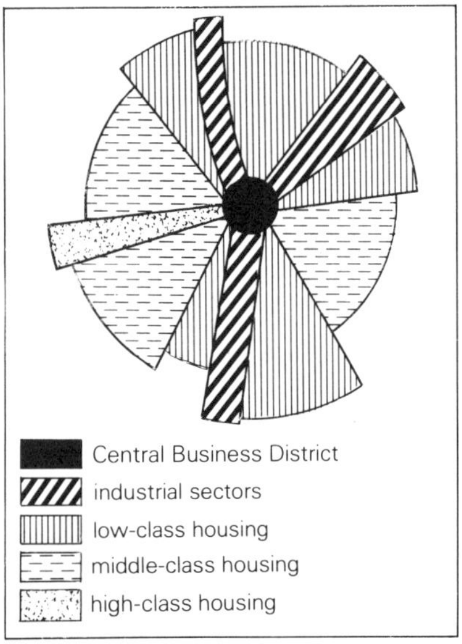

Figure 110 Sectors of an imaginary city

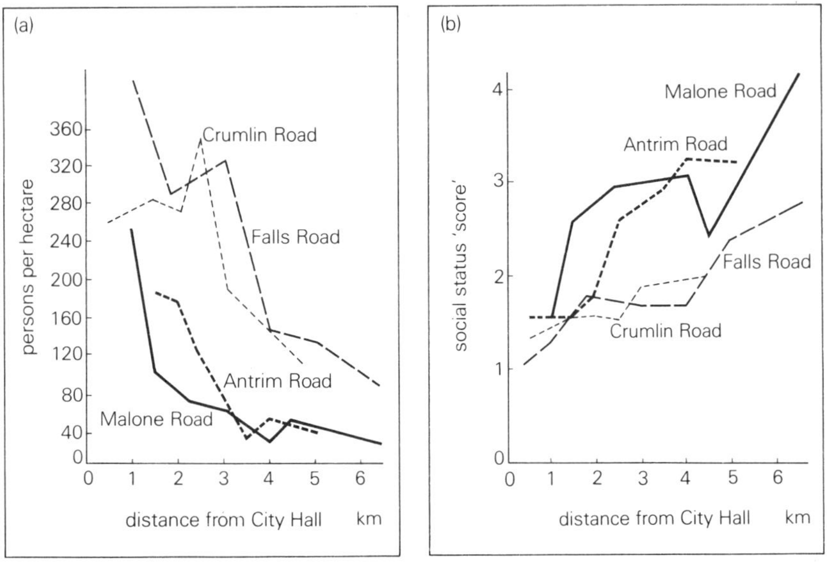

Figure 111 (a) Population density of Belfast (b) Social status measured along sectors in Belfast

status area, or a low one? Your answer to question 6 may help with this.

Density of population is closely linked to density of housing.

10 What do you think is meant by density of housing?
11 Look again at figure 108. Which photograph do you think shows the higher density of housing?

If you now look back to figure 107(a) you will see that the areas of different status are roughly arranged in **sectors**. If you have not come across this word before, it means, in this case, an arrangement like the one in figure 110.

As well as changing between sectors, status often gets higher as you go outwards from the city centre and population density goes down.

Earlier in this book you were asked to draw graphs. Now is your chance to try to **interpret** some. First, look back to figure 107 to see where the sectors are.

12 Look first at graph (a) in figure 111. For each street, is the density higher or lower at six km from the city hall than it is at one km?

13 All four lines show falling density as you go outwards from the city centre, even if in places there are slight rises. However, two of the roads have a lower density than the other two.
(a) Which two roads have generally higher density and which two lower?
(b) Which two roads are in higher-status areas and which two in lower-status ones? (Look back to figure 107.)
14 Now look at graph (b). What happens to the social status of the four roads as you go outwards? Give a reply about the general trend; ignore small changes.
15 Can you suggest any reasons for your answer to question 14?

If you now look back at figures 107, 108, 110 and 111, and at your answers to the questions, you will have a good idea of the **social structure** of Belfast. The social structure of most British cities can be identified in this way. There will usually be sectors and these nearly always decrease in population density as you go away from the centre.

Accessibility within the city

You may wonder why there should be a change in population density as you go outwards from the centre of a town. Like much else in geography, it has to do with accessibility.

In most towns, the most accessible part is the centre, where most main roads meet. This area becomes the main business centre, which is called the **Central Business District** or CBD, for short. Because the CBD is the most accessible, shopkeepers and office managers will want to have their premises there. This is because more people will be able to use the shops and there will thus be more profit.

If a great many shops and offices all wish to be in the same small area, there will be competition to buy, or rent, land. If there is competition, the rent increases. If there is enough competition, as there is in all UK cities, only the wealthiest shopowners will be able to afford a place in the centre. Usually this means the big chain stores, or the offices of national companies.

16 Even in the centre of strife-torn Belfast, you can see the same shops that are found anywhere in the UK. Which shops can you see in

Figure 112 Belfast city centre

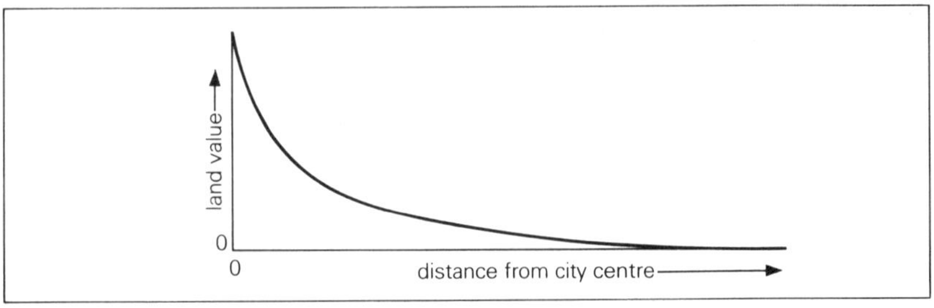

Figure 113 (above)
Figure 114 (below)

figure 112 which you would expect to see in the centre of any other city of the UK?

17 Now make a list of those shops which you cannot see in figure 112, but which are found in almost all UK city centres. The shops which you have listed belong to large companies, which is why they can afford the city centre rent. If the rents are so high, how can small shops like the one in figure 113 survive?

The answer is, they cannot. Even the well-known boutiques are usually part of a chain. Next time you buy clothes in such a shop, have a look at the back of the bill and you may well find a list of other shops in the group.

There are exceptions to this, of course, but independent small shops do not usually last long. They are either bought out by a bigger company, or they go out of business. If you go outwards from the centre, you will find that away from the principal shopping streets there are the smaller shops which cannot afford the highest rents. Even here, you will not find many houses. Shops generally can afford higher rents and rates than houses. Further out still there are houses, but rent for land is still high and blocks of flats are often found. As you go further out, the land prices, or rents, fall and so fewer houses are built on a given area of land.

Figure 114 shows a similar graph to figure 111, this time not for any particular street, but for streets in general. It shows that land value falls as you go further out from the CBD.

Imagine that a graph like this was cut out of card and stuck on to a map. The result would appear like figure 115.

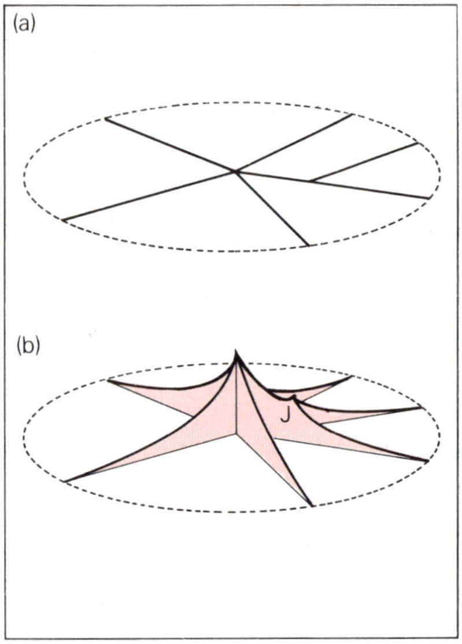

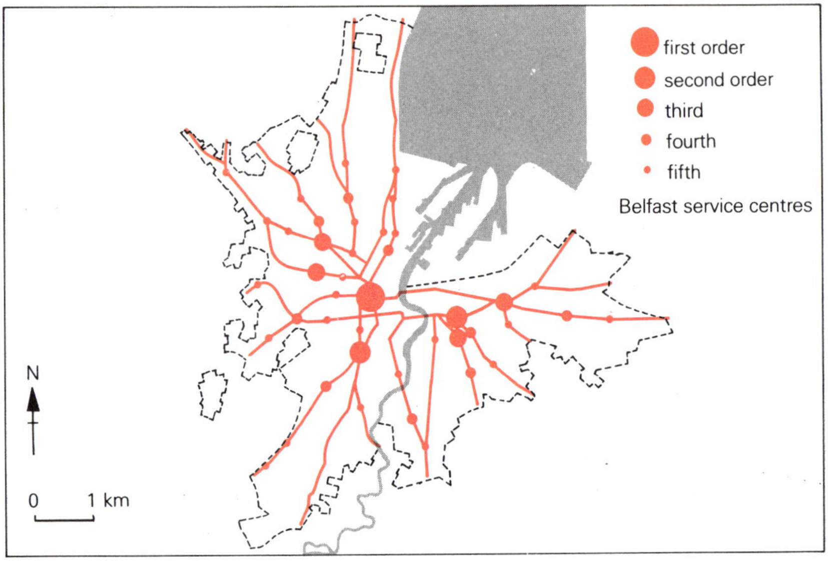

Figure 115 (above, left)   (a) Road pattern of an imaginary city
(b) Land values in an imaginary city
Figure 116 (above, right)   Belfast: shopping centre hierarchy
Figure 117 (left)

---

18 Why does the shape change at the point marked J?

This arrangement is a simple one. The city centre is obvious and there is only one smaller centre at point J. In Belfast, or any other town, there are many other smaller centres. Each of the smaller centres is less accessible than the main CBD so the land values are not so high. In the smaller centres, there are not so many shops and they are of a more local kind. People travel to these more often but do not travel so far. In places that are still less accessible, there are even smaller centres. In the same way, there are centres of gradually smaller size, at gradually less accessible places. The very smallest 'centre' is a corner shop, or small shopping parade, which people visit daily for bread, confectionery, newspapers etc.

19 Look back to pages 70 and 71. Explain in your own words the way in which service centres inside cities are like village, town and city centres.

This kind of **order** in shopping centres is called a **hierarchy**, which means that the centres have 'ranks' almost like the ranks in the Army. Figure 116 shows the hierarchy of shopping centres in Belfast.

20 Which of the shopping centres in figure 116 is the CBD?

Belfast is a fairly simple example of a hierarchy. What about the conurbations? In these, the arrangement of shopping and commercial centres, of land prices and population densities is much more complicated. This is because there are so many 'town centres' within the conurbation. However, the way the hierarchy works is the same as in Belfast or any other city.

You may remember that on page 81 we said there were no houses near the CBD and that only big stores and offices could afford the rents. This competition for land has given rise to yet another form of disagreement over land use.

21 Can you work out what the message on the banner in figure 117 is likely to say?

This sort of argument is going on in many towns. Where a town centre is being redeveloped, the plans often include office building. This is because offices can afford town centre rents, and they bring in more rates than houses.

Residents of the town often feel that this solution is making the housing shortage worse. Most towns have a housing shortage. The offices will need parking places for the workers' cars and these cars themselves will make traffic congestion worse. Townspeople are concerned that the centre will become 'dehumanised' by having too many offices. The name 'concrete jungle' sums up many people's fears.

What would you do if you were on your local council?

Functional zones

Whilst looking at the geographical arrangement inside the city, we have mentioned two or three types of building. In particular, houses, shops and offices have been named. We have also implied that the houses are in one part of the town, a **residential area**, and that the biggest group of shops and offices is in another part, the CBD. Are there any other **zones** in which there are groups of buildings with the same **functions**?

In many towns, residential, commercial (that is, shops and offices) and industrial functional zones can be seen. Industrial zones are particularly important in the industrial towns of the UK. Do you know which towns are industrial? Deciding exactly what is an industrial town is quite difficult. One way is shown in the exercise that follows.

22 From figure 118 make a list of those towns with more than the national average employed in industry. This is a way of defining an industrial town. Then make a tracing of the UK from an atlas, showing the towns you have listed.

23 The method given in question 22 is a very rough one. Using the figures in figure 118, can you think of a better way of defining an industrial town?

Industrial towns are fairly well-known as a type of town, but are there any other kinds? Traditionally, there are supposed to be market towns; dormitory towns; administrative towns; resorts and ports, but it is easier to name them than it is to identify them.

24 Do you think there is likely to be a functional zone containing holiday activities in a resort town?

Market towns certainly exist and can be defined as those places which have a market. However, in most such towns some other function is now far more important than the market.

Cities in rank order	Total employed persons	% in manufacturing industry
Greater London	4085530	27
Birmingham	586490	50
Manchester	335540	32
Liverpool	325200	31
Sheffield	266300	46
Leeds	255700	37
Bristol	215580	32
Teesside	182700	48
Leicester	177500	49
Coventry	177100	60
Nottingham	174640	37
Newcastle-upon-Tyne	152710	28
Stoke-on-Trent	146920	50
Bradford	145140	42
Cardiff	144390	24
Wolverhampton	136510	53
Hull	132890	32
Southampton	112450	26
Derby	111300	48
Portsmouth	110990	29
Plymouth	107230	28
National average		34

Figure 118 Large towns of the UK: proportions employed in manufacturing industry. Source: Census

Dormitory towns are those where most of the population work in some nearby large town and commute between home and work. They are much the same thing as commuter villages. Most of our large cities have a 'commuter belt' around them, and the bigger the city, the wider the belt.

25 What do you think will be the main functional zone in a dormitory town?

26 Look back to figure 96 which shows the main commuting areas. If you live near any of the main employment centres, try to identify commuter villages from the one-inch or 1:50000 map.

27 When you think you have found the dormitory towns, try to visit one to see if you were right about its main function. What investigations would you carry out when you got there to check your guess?

Another kind of town is sometimes called 'administrative'. This means a place where government, either local or national, is important. Usually such towns are so big that they have many other important functions as well. However, there is often a distinct functional zone of government offices.

The most important administrative city of the UK is also the most important industrial city and the most important residential city; in fact, generally the most important. It does not take much to guess which city this is.

28 Look at figure 119. What are Downing Street and the Houses of Parliament famous for?

Many of the buildings marked are concerned with national government. You will see what a large proportion of the space they take up. Despite London's many

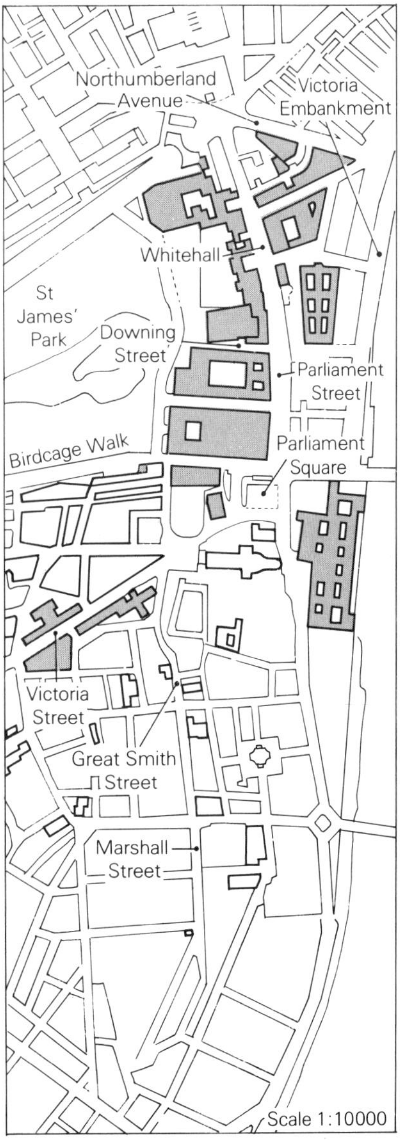

Figure 119

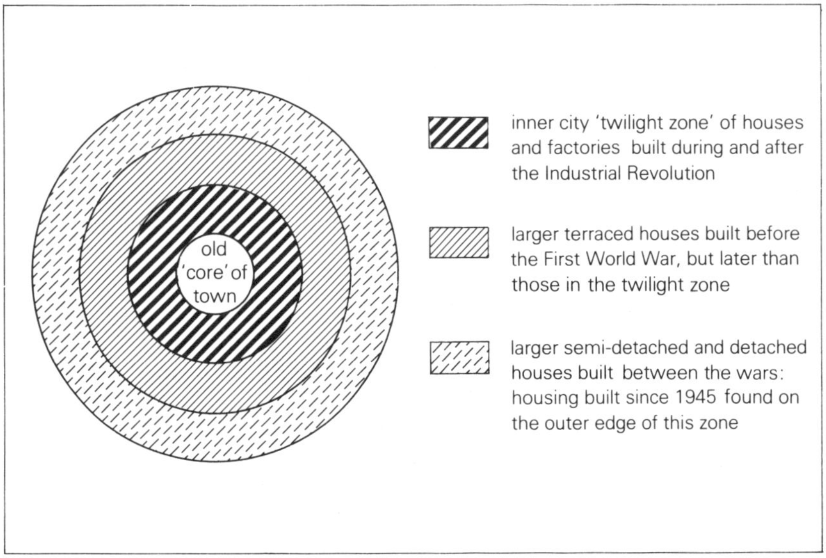

Figure 120

other functions, there is a very clear administrative zone. Another functional zone that is fairly easy to spot is the docks area of a large port, though it is often very difficult to separate this from an industrial zone.

To sum up, all towns have residential zones and service zones. Some also have industrial zones which can be identified on a map. Of these, some are of the specialised port industry type. Smaller numbers of towns have administrative zones, and smaller numbers still have entertainment or holiday zones, such as the 'Golden Mile' at Blackpool.

Age zones

Another type of zone is shown in figure 120. This is a 'model' and is not supposed to be the way a real town would look.

You may think it is obvious, since it simply shows the town growing outwards from the centre. Like the models we have met earlier, its purpose is as a starting point for study. As you can see, it has no roads, no railways and none of the sectors we mentioned earlier.

Figure 121 is also a model, but on this one almost all the features can be recognised in almost all towns of the UK.

29 Is map (d) in figure 121 in any way like figure 120? In figure 121, map (a) shows the old core of the town, situated at an accessible point before the Industrial Revolution. With the arrival of the railway (map (b)), and industrialisation, more people were needed to work in factories and they could now travel much further to work. The factories and workers' houses spread over a very large area. Between the two world wars (map (c)) came the real beginning of the motor age. People could travel even further to work and did not have to be near a station. New roads were built and houses and factories 'sprawled' out along these in the form of ribbon development. Since the Second World War there has been a much greater increase in the use of the car. At the same time, planning policies have been introduced aimed at stopping massive sprawl. The unplanned growth of towns before the Second World War was threatening to eat up all the countryside. One result of planning controls together with the ability to travel long distances to work, has been the growth of commuter villages.

The age zones shown in figure 121 are easily found in most towns. It is true that in the bigger industrial cities, most of the core was rebuilt in Victorian times and may not be easy to spot, but there are usually a few old buildings left. Present-day rebuilding of town centres may also make the old core hard to find, but there are usually some traces of it. One of the projects at the end of this chapter suggests that you try to draw the age zones of your own town.

Figure 121 (page opposite) (a) the pre-industrial 'core'
(b) the Victorian and Edwardian age
(c) the inter-war years
(d) the age of planning

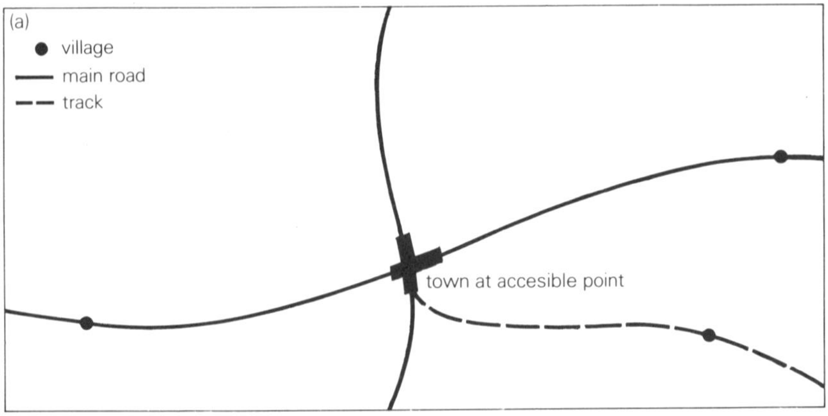
(a)
village
main road
track
town at accesible point

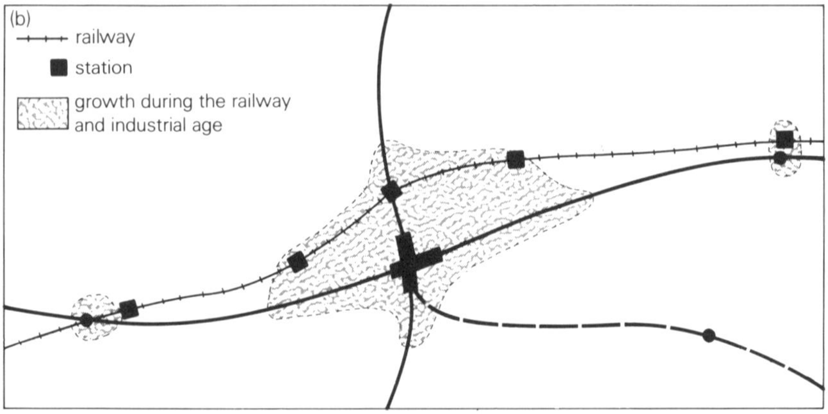
(b)
railway
station
growth during the railway and industrial age

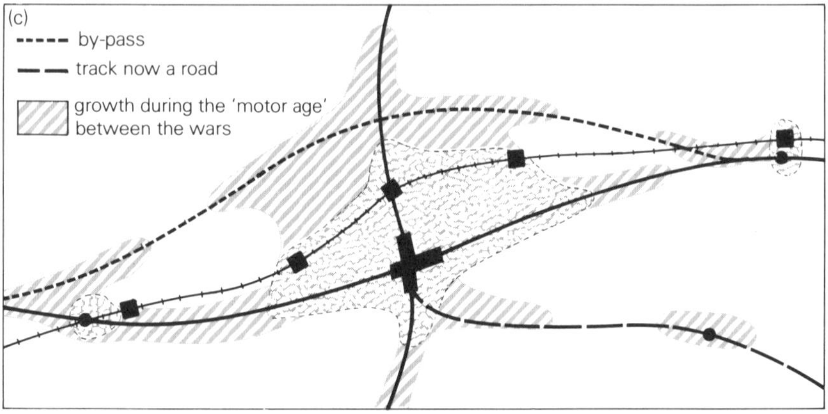
(c)
by-pass
track now a road
growth during the 'motor age' between the wars

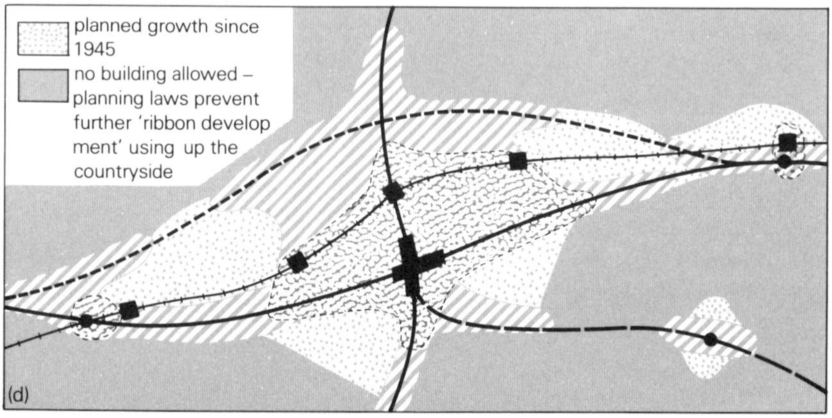
planned growth since 1945
no building allowed – planning laws prevent further 'ribbon develop ment' using up the countryside
(d)

CONCLUSION

The various zones dealt with in the second half of this chapter have concerned mainly buildings and land. yet towns are not just bricks and mortar, they are collections of people. It is the behaviour of people, their social organisation and their conflicts that makes the urban geography of the UK most relevant to you. That is why the main part of this chapter has been given over to attempting to understand these activities.

PROJECTS

1 You can draw a map to show the population density of your own, or your nearest town.

You will need the county report of the census and a map of your town showing the electoral wards, as they were in 1971.

The population density is shown in the census report for each of the electoral wards. You simply have to devise a colour scheme, and colour in each ward.

A suggested colour scheme is:
0 to 10 persons per hectare — yellow
10·1 to 20 pph — green
20·1 to 30 pph — light brown
30·1 to 40 pph — red
over 40 pph — black

Does your pattern of population densities fall into sectors? Is the density higher nearer the centre than at the edge of the town?

If neither of these has occurred, there could be one simple reason for it.

You may know of an area in your town like figure 122. The houses are closely packed and so the population density ought to be high. However, in the census the density is worked out by dividing the total population by the total area of the ward. (Look back to page 79 and question 8.)

A better way would be to measure the area occupied by houses only. Then you could divide the population by this number. The answer would be the **residential** population density, which is much more useful than the one in the census.

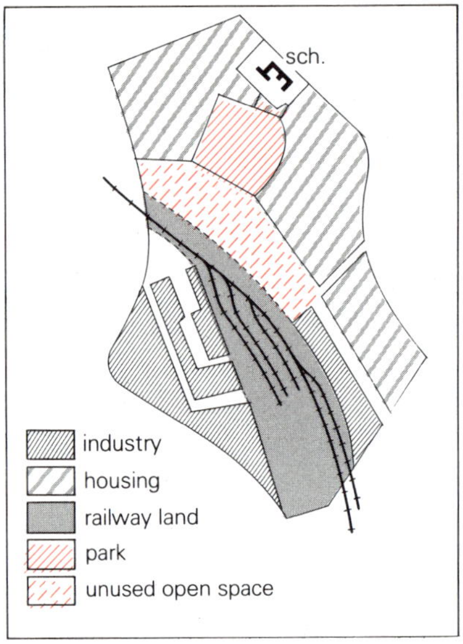

Figure 122 Imaginary electoral ward

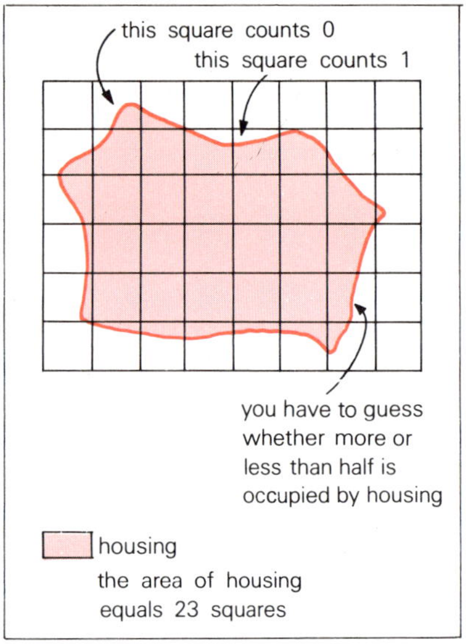

Figure 123 An example of measuring area by counting squares

You can measure the area of housing by placing tracing graph paper on top of the electoral wards map and counting the squares covering housing. At the edges of the ward, you count a square as a single whole one if more than half is housing, or nought if less than half of it is occupied by housing.

When you have counted the squares, you can convert them to hectares, according to the scale of the map you are using.

Now divide the population by the number of hectares to obtain the population per *residential* hectare for each ward. Devise a colour scheme and draw a map as in the first part of the exercise. Remember to colour in the whole of the ward, even though you are only dealing with the residential part of it.

If you do this kind of map, it is more likely to show up sectors and decreasing densities as you go away from the centre. You can try it using figure 109, which has the information needed.

2 Conduct a survey to find the most accessible place in your town. This needs a large group, perhaps a whole class, to do it properly. First, make a base map of the town centre, by tracing the roads from a 1:10000 or 1:2500 map.

Station people around the town centre to count the number of pedestrians. The more counting places you have, the better. Try to avoid counters being stationed near bus stops as these produce sudden changes in the pedestrian flow.

At a given starting time, each person counts the number of pedestrians who pass him during, say, a ten-minute period. The best way is to have counters on each side of the road being responsible for their side only. Pedestrians crossing the road do not count. If part of your town has been 'pedestrianised' the simplest method is to have two people standing back to back in the centre of the street.

When you have finished the pedestrian count, add to your base map the department stores and the chain stores in the town centre (e.g. Woolworth's; Boots; W. H. Smith; Marks and Spencer).

Then, back at school, draw a

pedestrian flow map. See if there is any one point which you could call the most accessible to pedestrians. Measure the distance of each department or chain store from this point. Name the roads which have the highest pedestrian flow. Are there any of the big stores which do not have at least part of their frontage on at least one of those streets?

What do you think is the purpose of the last two parts of this exercise?

3 Investigate the shopping centre hierarchy of your town. You cannot visit every shopping centre, so select three or four that you think are of different size. For each centre:

(a) map the roads entering it, making clear the class of the road
(b) conduct a pedestrian count, as for the previous exercise
(c) make a list of the types of shop in each centre.

Then try to arrange the centres in rank order. You could rank them according to:

(i) total number of shops
(ii) total number of pedestrians counted
(iii) accessibility score of the roads entering it.

For each rank, you could see if there are any types of shop which are not found in the rank below. Shops that are found only in the higher ranks would need to be in the more accessible place so that more people could use them. For example, you might find a department store only in the city centre; a jewellers only in the city centre and the next rank down; a ladies' shoe shop only in the city centre and the second and third rank; a hardware shop only in the top four ranks, and so on.

8 Knock it all down and start again

Homes fit to live in

1 Look at the pictures in figure 125. Does Hattersley look old and dirty? Look at the open country beyond. Does that look attractive?

Many people think Hattersley is a clean, pleasant place, set in beautiful countryside. Why, then, should the singer wish he 'wern't 'ere'? Another part of the song goes:

'In our toilet in the backyard
You could sit and watch the stars'

but if you look at figure 126 you will see what these toilets, and the houses they belonged to, were like. It might make a funny line in a song, but do you think that anyone would really prefer an outside lavatory to an indoor one? The houses themselves

Figure 124 'The Hattersley Lament'
Figure 125 (below, left) Hattersley
Figure 126 (below, right)

Figure 127

are old, insanitary, dark and small.

Why then, should there be people living in Hattersley who did not want to go and live out in the country? It may have something to do with a sense of community. Planners are now beginning to realise the great importance of a sense of community.

In the crowded slum areas, despite the poor houses, everybody knew everyone else. They always met people they knew as they went about their business or visited the shops.

2 In a residential area which is mostly 'grid iron' pattern streets, where do you think the most accessible places are?

3 How many shops can you see in figure 127? In what sort of place is each one located?

In addition to this neighbourhood loyalty, people were used to living near the city centre. They were used to an urban way of life.

The 'redevelopment areas' on figure 128(a) are the places where there is most 'substandard' housing.

4 How far is the furthest point of Gorton away from the city centre?

5 How far is Hattersley away from the city centre?

In Hattersley the residents were given something very different from what they knew. The comparatively

Figure 128 (a) Principal Manchester redevelopment areas
(b) Manchester and Hattersley

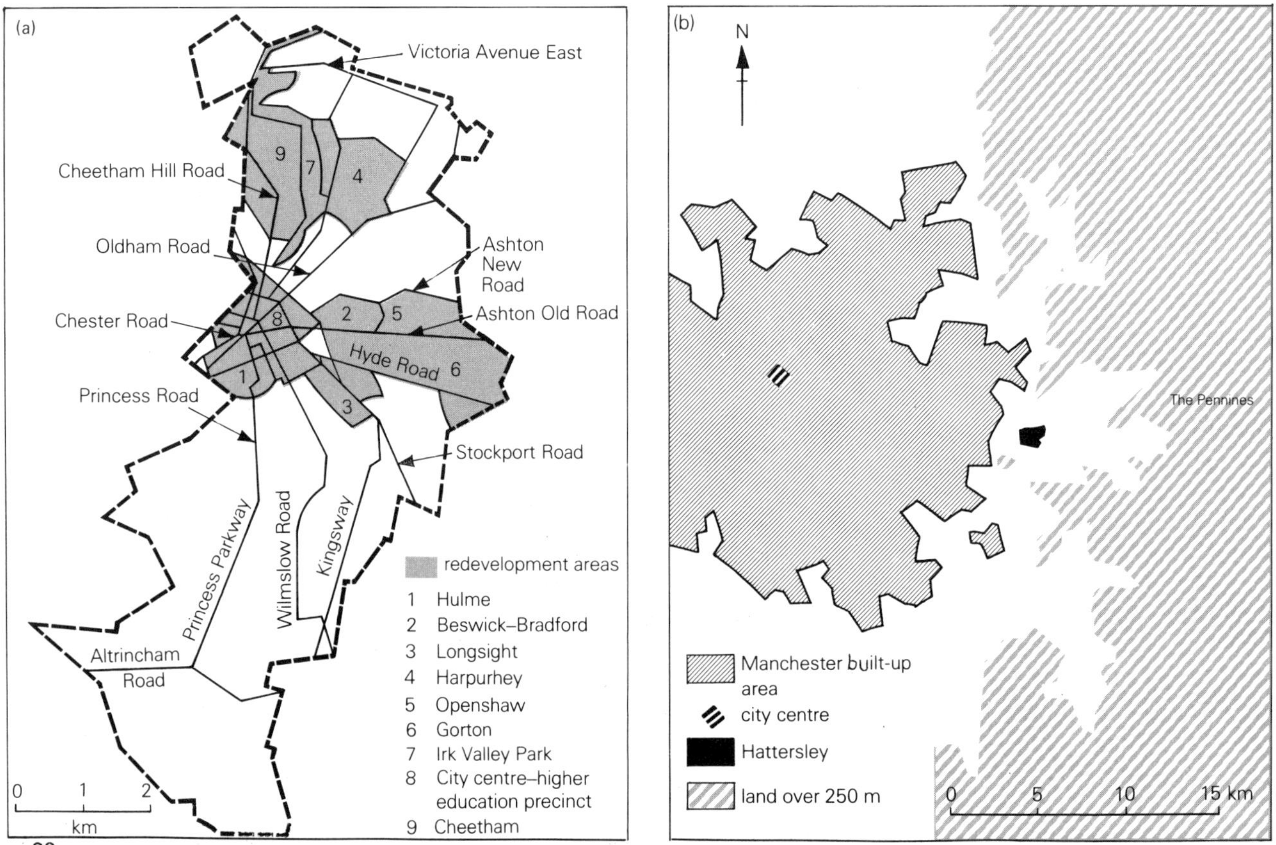

large gardens, the wide streets with grass verges, the grassy open spaces, give the people *too much* space and make them feel uncomfortable.

Instead of being able to nip down to the corner shop for daily needs, a major shopping trip to the Hattersley centre is now necessary. For the attractions of a big city, it is now a long journey into the centre of Manchester. The nearby centre of Hyde can provide some of the more important services, but, as you know from the previous chapter, it cannot compete with the centre of the conurbation.

So far, everything about Hattersley seems to be wrong. How did it come to be built in its present location and style? An official Manchester booklet reads:

> . . . the eradication of Manchester's legacy of ill-planned and unfit dwellings and the rehousing of the displaced people.

6 Make sure you know what this passage means. What is 'eradication'? What is a 'legacy'? how does Manchester come to have a 'legacy of ill-planned . . .', etc. housing?

Like most large UK cities, particularly those which grew during the Industrial Revolution, Manchester has a 'collar' of poor housing in the inner area, just outside the city centre. (Compare this description with figure 120.) In these areas, poor housing and industry were all mixed in together, perhaps in sectors near the canals and railway lines.

7 Would you expect these houses to be built at high or at low densities? Remember they are near the city centre.

The houses were also poor quality. The usual solution is to call them 'redevelopment areas', to knock

Figure 129 Redevelopment (a) The beginning (b) The site is flattened (c) The result

down all the old buildings and to replace them with new ones. At the same time, new roads, schools and sometimes offices and factories can be built.

Now look at the figures in figure 130.

Figure 130 Estimated population before and after redevelopment in the first five of Manchester's redevelopment areas

	Before	After
1 Hulme	24 500	12 600
2 Beswick	23 000	12 500
3 Longsight	20 000	11 000
4 Harpurhey	15 000	10 700
5 Openshaw	15 000	8000

8 Altogether, how many more people were there before redevelopment than after?

9 What do you think happens to these people?

When new houses, flats and maisonettes are built, much more open space is left. This may be for children to play in, or just because the planners think it looks nice. New roads are wider and new schools are built, all taking up space. Because of all this, there are people whose houses were knocked down but who are not given a new house in the same area. It is these people who are moved to places like Hattersley, which are sometimes called 'overspill areas'.

Satellite towns

All the big cities of the UK have such overspill estates. In addition, in many cases there are so many people to be rehoused that there is not enough room on the edge of the city. In any case, if the city went on building estates round the edge, there would soon be no open country left.

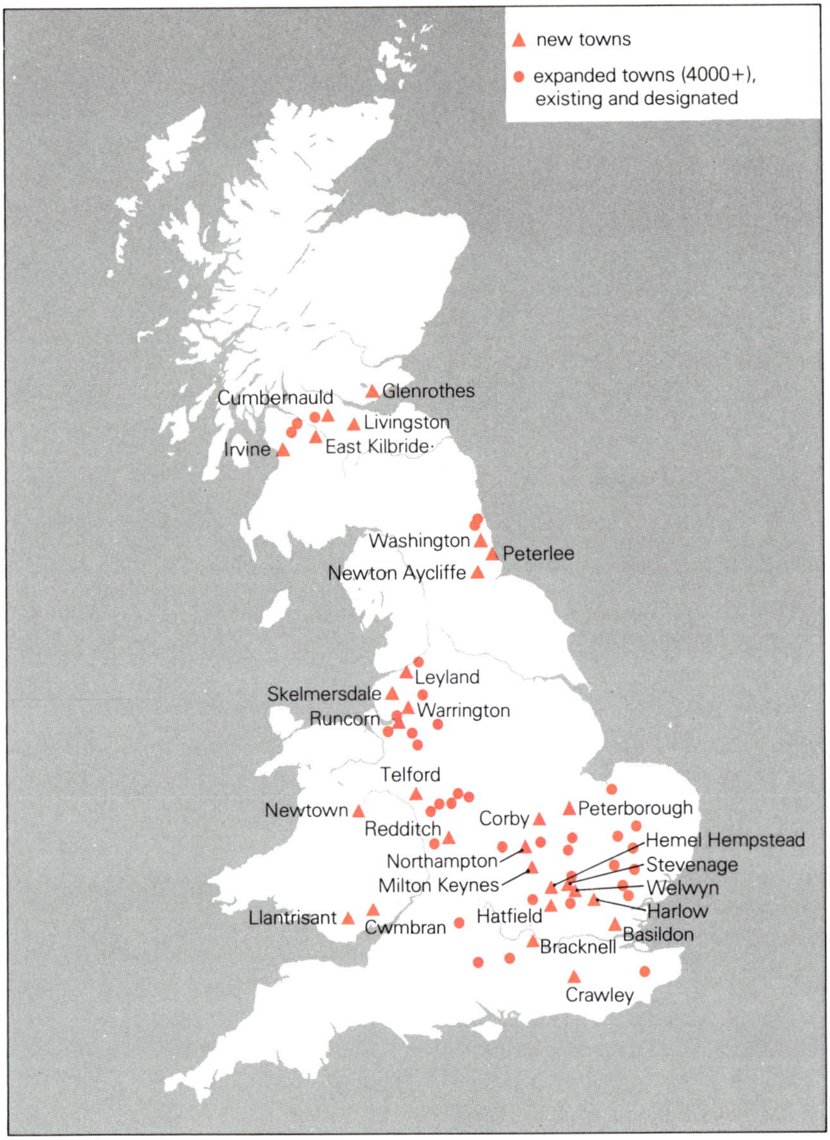

Figure 131 Expanded towns (over 4000) and New Towns

One way in which many cities have tried to deal with this is to make arrangements with other towns to take some of the extra people. The 'parent' towns help to pay the cost of building new houses and of encouraging industries to go to the 'satellite' towns. Manchester has four such 'expanded town' agreements.

Although these places are now part of the Metropolitan County of Greater Manchester, they were separate when the agreements were made. Manchester has four agreements, but other cities have more. London has 31, Birmingham has 15 and Glasgow has 47 (figure 131).

10 How many expanded towns are there with intended populations of over 4000?

There are more people in need of homes than can be provided by the expanding towns. To house these, entirely new towns have been built since the Second World War. It is now time to look at why new towns were needed, apart from coping with overspill.

Urban sprawl

In the 1920s and 1930s, when the use of the motor car and bus was increasing rapidly, there was a vast expansion round the edges of our towns. There were plenty of people who wanted to move out of the town centre to where they could have a garden and a bit more space. There were many others who came from the country to work in the town but who wanted to be away from the grime of the city centre and nearer the open country.

11 Do you notice anything odd about these people's wishes? What was Hattersley criticised for in the earlier part of this chapter?

At first most of the houses were built along the main roads, many of which had to be enlarged because of increasing traffic. Then the spaces between the main roads filled up. There were not only houses; space-consuming factories were built too.

It all added up to what has become known as urban sprawl.

Green belts

After the Second World War a 'green belt' was established around London, where the sprawl was worst. The purpose of the green belt was to stop further expansion and to preserve some open country within reach of the city. To achieve this, almost all building within the green belt was prohibited. The green belt was thought to be such a good idea that now most of the big cities and conurbations have green belts planned or in operation.

In the case of London, people were tempted to live beyond the green belt and to work in London. This created

Figure 132 London's Green Belt and ring of New Towns

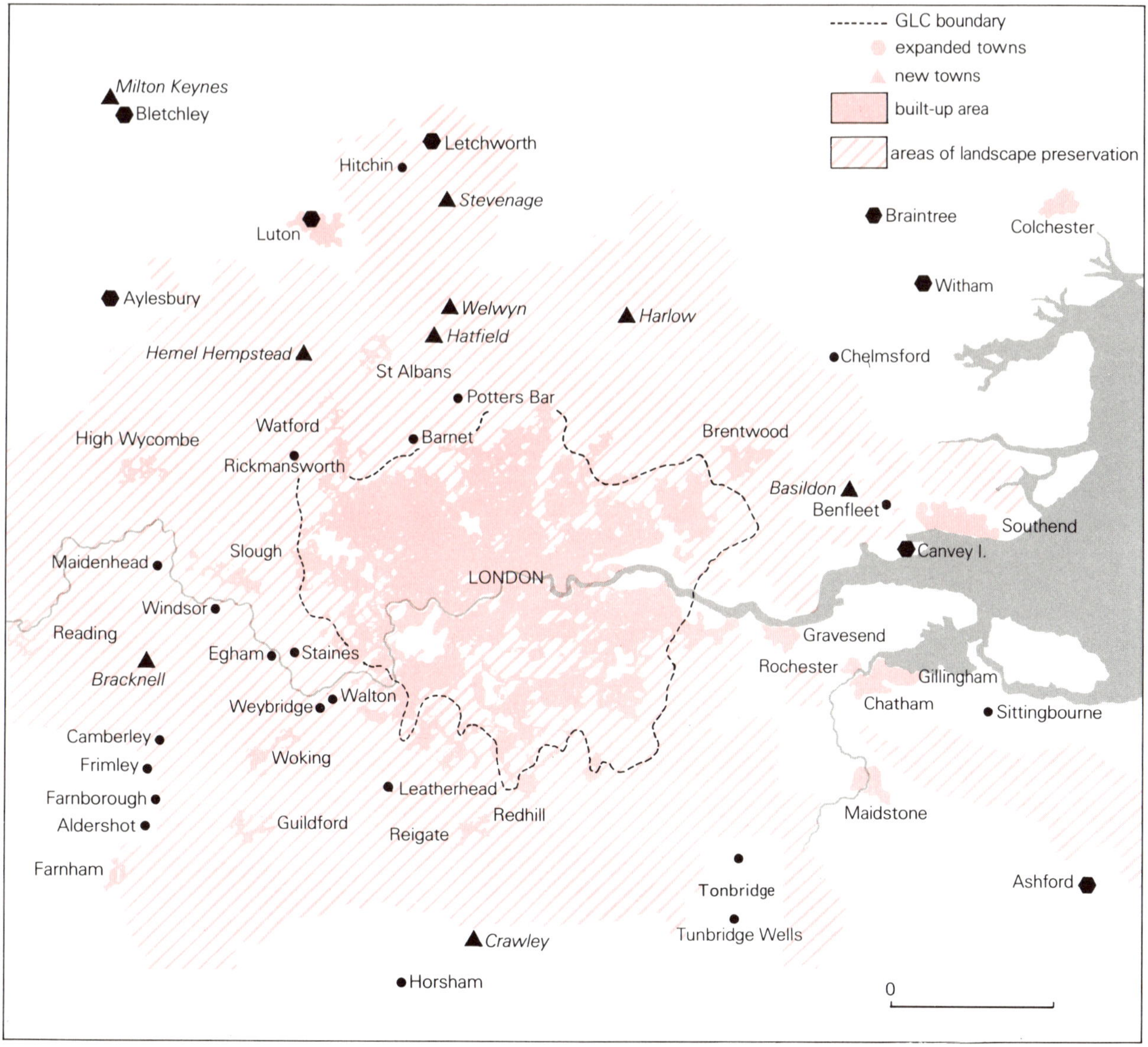

the commuter towns and villages which we have already seen (pages 71 and 83) and it also caused considerable overcrowding on London's transport systems.

We thus have several problems; first, overspill for people from redevelopment areas; second, urban sprawl which was threatening to happen again in the commuter towns; third, overcrowding of London's transport; fourth, the general housing shortage, particularly in South East England.

The idea of the new towns was that they should have employment and shops as well as homes. The people who lived in them would not need to travel outside the new town for anything except major shopping goods. At the same time, they would be near to London, so that people from London's redevelopment areas would not feel too far from 'home'.

12 Look at figure 132. How far is it to the furthest new town from central London?

13 How many new towns are in the 'London ring'?

In addition to attempting to solve the four problems mentioned earlier, the new towns could be designed to avoid some of the difficulties of older towns.

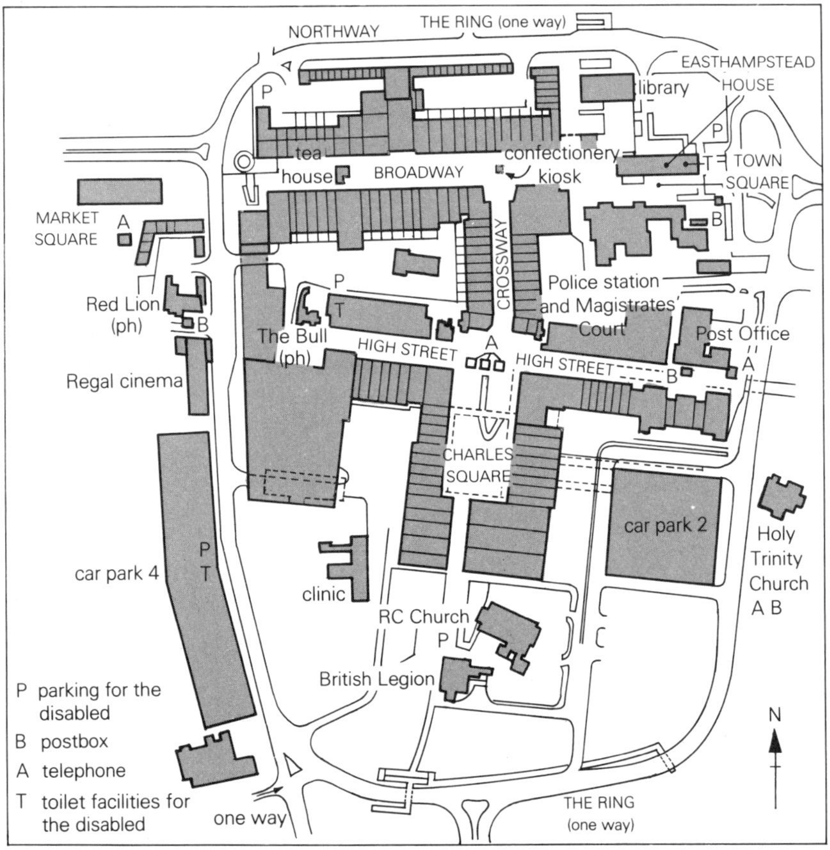

Figure 133 (above) Bracknell town centre
Figure 134 (below) Bracknell town centre

Traffic and towns

Most of the centres and inner areas of our cities were built before the age of the motor vehicle. As cars increased in numbers these areas had to be adapted to suit them. Anything which is made for a different purpose and then altered is rarely satisfactory.

But the new towns could be built with the motor car in mind.

14 Before reading on, can you

Figure 135 Housing in Bracknell

think of any ideas which could be used but which are not found in older towns? Make a list of your suggestions and compare them with others in your group.

15 Compare figures 133 and 134. On the photograph find:

(a) The Ring
(b) Car Park 2
(c) Charles Square
(d) Crossway
(e) High Street
(f) Broadway.

The last two are more difficult because you can only see part of them.

One of the basic ideas is to *separate* cars and pedestrians. A main ring road takes traffic round the centre, from which you can get to car parks. The car parks are within easy walking distance of the shopping centre. In Bracknell they are connected to the shopping areas by covered walkways.

If you have 'pedestrians only' in the main shopping streets, how are the goods delivered to the shops? To do this, there are roads leading from the ring to the *backs* of the shops. These are for delivery vehicles only.

16 In figure 133 the main shopping areas are Charles Square, High Street, Broadway, and Crossway. Look carefully at the photograph, with a magnifying glass if needed, and see if you can see any pedestrians. There should be no cars!

17 Find the covered walkways from the car parks to the shopping areas.

18 Find the rear access areas for delivery vans and official vehicles.

Apart from the centre, the new towns gave an opportunity to improve the residential areas. These could be designed to allow residents to use their cars but at the same time prevent through traffic.

19 In figure 135 look carefully at the road pattern in the housing area. Like the town centre, there is a ring road around the outside, which connects with the main road system. Leading off the ring road are smaller ones running in among the houses. Trace one of these with a finger and see what happens to it. Are there any roads *through* the housing area? (The whiter strips are footpaths, not roads.)

20 Can you see any garages and parking places?

You can also see in figure 135 that there is a good deal of open space among the houses, as well as the playing field on the right. The industrial estates at Bracknell and other new towns are well away from the housing areas, yet they are connected to them by footpaths and roads. They also have good access to a main road.

21 You have now seen several

pictures of Bracknell. Would you like to live there?

Whatever your answer, give reasons for it.

Problems of New Towns

Many of the inhabitants of New Towns are not entirely happy. One of the reasons is that there are now far more cars than there were when Bracknell was planned. The residential areas are overcrowded with parked cars and the roads between the industrial areas and the workers' homes are choked with traffic at rush hour. Later housing in Bracknell (the town is not yet finished) is taking account of these drawbacks, but the earlier parts cannot be changed.

Cumbernauld is one of the later New Towns. More effort was made from the start to come to terms with the motor car. Part of one of the town's brochures reads:

> . . . provision is being made at Cumbernauld for one garage or parking space for every house, together with visitors' parking in housing areas and parking for some 6000 cars in the . . . town centre.

People and traffic are separated in other parts of Cumbernauld besides the town centre. The children in figure 136 never have to cross a main road on the way to school, nor need the postman when he is delivering his letters. All main roads are provided with underpasses such as these.

The road system keeps local traffic separate from the main roads. Where local roads join main ones, there are 'clover leaf' junctions, allowing traffic to filter in.

As the brochure says:

> There are no traffic lights, traffic wardens, policemen on point duty, and there never will be any need for them in the foreseeable future.

Is everything perfect, then, in Cumbernauld? Well, it seems not. Look at figure 137.

The houses, as you can see, are on a hillside. This gives them a good view.

22 Look up in an atlas the positions of Cumbernauld, the Campsie Fells and the Kilsyth Hills.

In fact, most of the houses of Cumbernauld are built on hillsides, with the town centre on the hill top. 'Whoever would be daft enough to design a town on a hill top in Scotland?' the local people are heard to say. Yet there *is* a very good view, and the walk back from the shops (with full shopping bags) is downhill.

The hill top is also very windy. There is open space all round the town centre. In addition, there are not enough shops.

These last two problems are inevitable in a town that is genuinely new. The shops and other town centre services cannot be provided until there are enough people to use them. Cumbernauld has not yet reached its planned population. In Cumbernauld, all that open space will eventually be filled by the growing shopping centre and other services.

Will the New Towns ever be big enough to provide all the services that city dwellers are used to? The answer is probably not. In an attempt to meet this difficulty some of the newer New Towns are intended to be larger still.

In particular, it was hoped that the trio of Milton Keynes, Northampton and Peterborough would between them provide all the facilities of a really big city.

23 Look back to figure 131 to see how far apart these three towns are.

Figure 136

Figure 137 'Split-level, all-electric terraced dwellings. These houses . . . are all sited to capture the view to the Campsie Fells and Kilsyth Hills while maintaining individual privacy.'

Figure 138 Kildrum neighbourhood unit, Cumbernauld

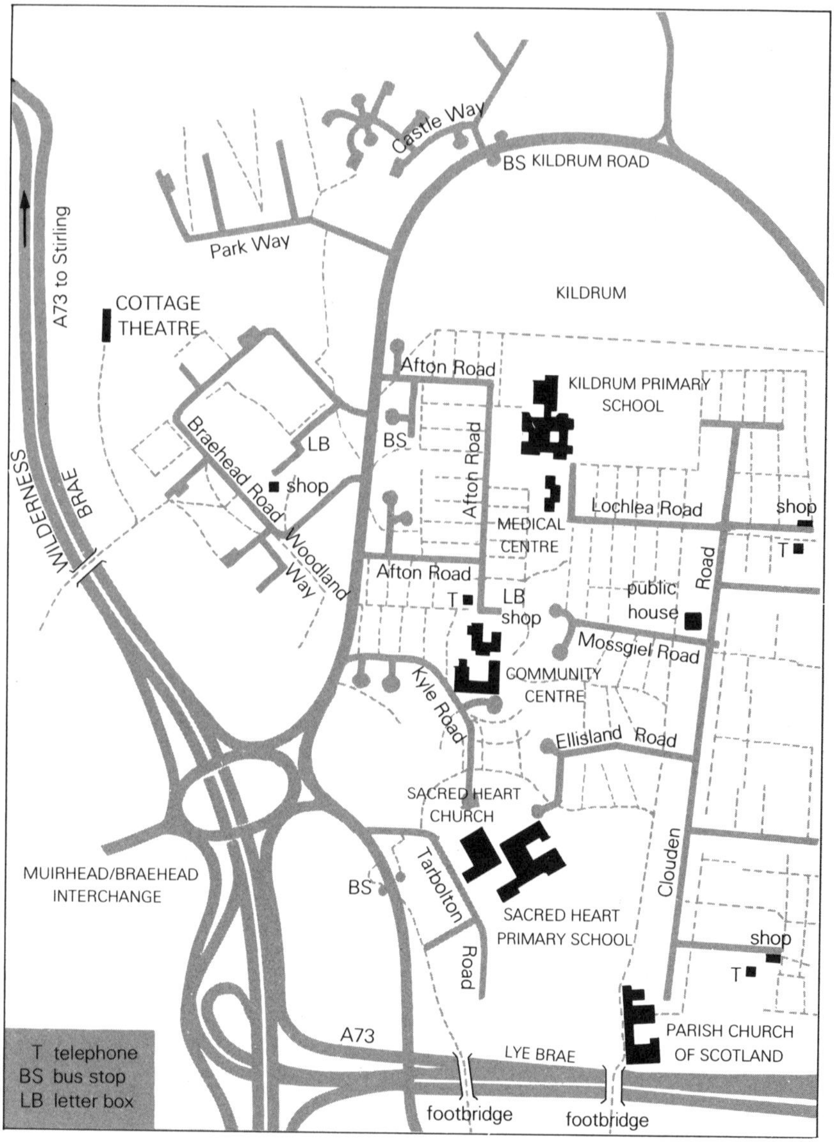

Other difficulties of New Towns are similar to those expressed by the people of Hattersley. There is a feeling of being cut off from the place you were brought up in and, most of all, a loss of the sense of community. All the new towns' residential areas were designed as 'neighbourhood units', to try to create communities.

24 Figure 138 is a plan of a neighbourhood unit. What services are provided within the area?

25 Are there any routes which could be used by through traffic?

Despite the provisions for community life which you have seen, it appears that more than bricks and mortar are needed. People still feel that they do not 'belong'.

Although there are difficulties, there are many who are well satisfied with new town life. Mostly these are young married couples with small children, who welcome the extra space available and the open country nearby. (Remember Hattersley!)

As for the problems, perhaps time will cure them. More and more services will be provided as the towns grow, and perhaps a sense of community will develop as children grow up in the New Towns.

Traffic and towns again: Birmingham's new city centre

If large-scale redevelopment is carried out in a city centre, there is an opportunity at the same time to try to solve the city's traffic problems. Different towns approach this task in different ways, according to their local means and needs. One of the biggest town centre schemes in Britain has taken place in Birmingham.

Figure 139

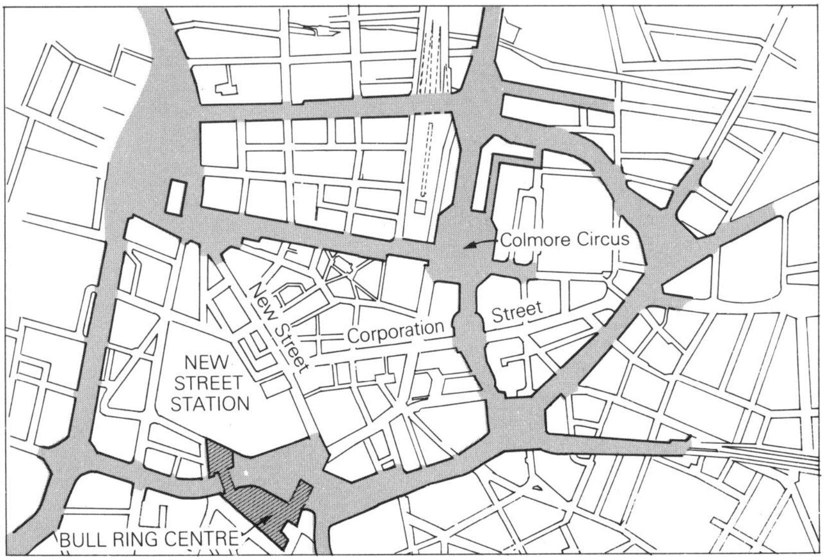

The underpass and junction in figure 139 are part of the massive inner ring road system, which is shown in figure 140.

This great road system encircles Birmingham's CBD and roads leading to Colmore Circus cut right into the heart of it. The roads leading off the Ringway link up with the radial routes which focus on Birmingham and which make the city one of the most accessible in the UK (figure 141).

Figure 140 (left) ***Birmingham's inner ring road scheme***

Figure 141 (below) ***Birmingham as a route focus***

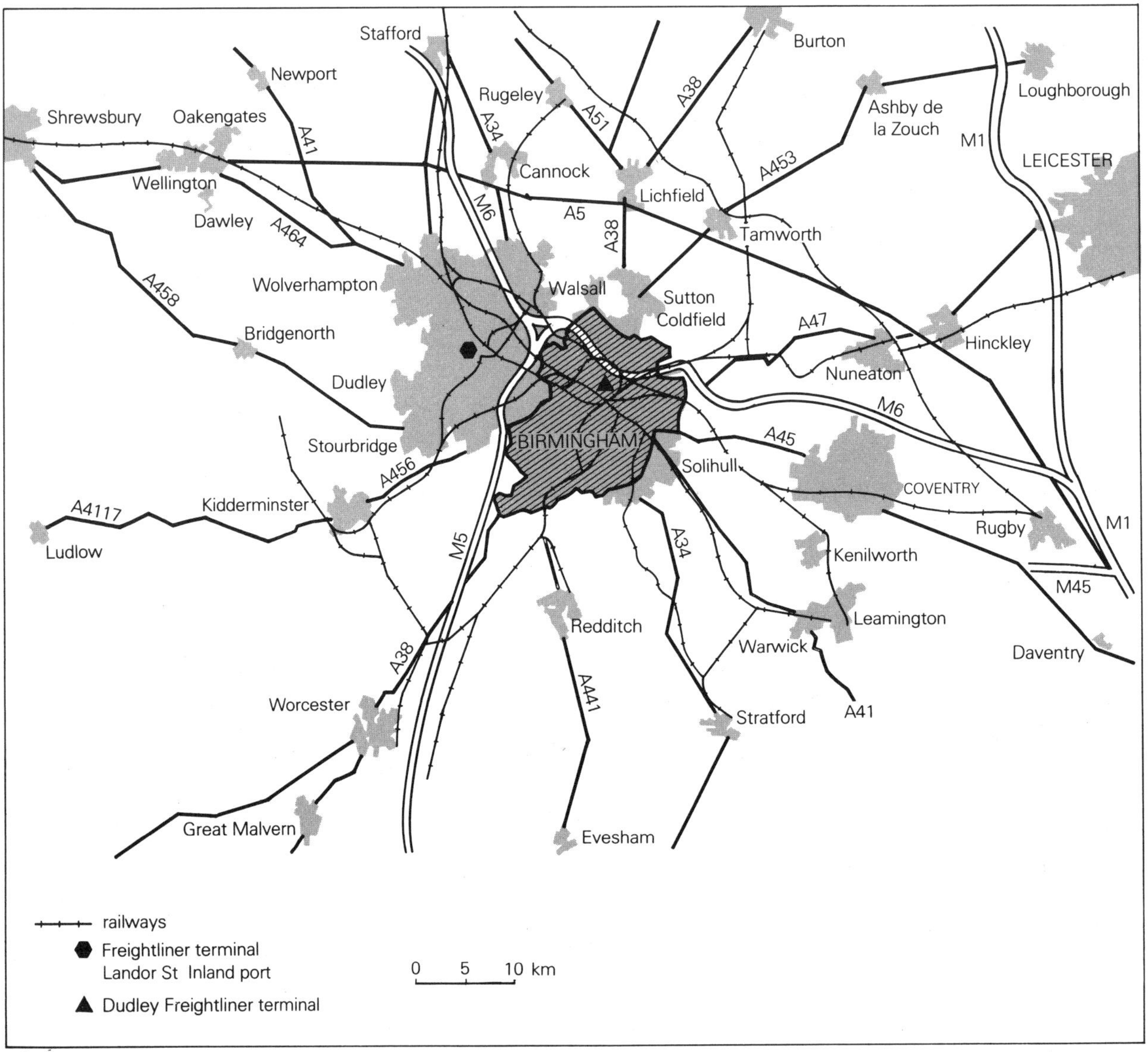

On all but one of the huge roundabouts that link up with the radials, there are underpasses to allow free progress round the ring.

This new road system cost many millions of pounds. It would clearly not have been worth building it unless it was linked with other redevelopment. Many of the new buildings are office blocks, like those in figure 139, but there are complex, multi-level shopping centres as well. The Bull Ring (figure 142) is the biggest.

In all this, the pedestrian has not been forgotten. The shopping centres like the Bull Ring are completely separated from the traffic, yet they are very near to, or over the top of, bus stops and car parks.

26 Can you see the entrance to the car park under the Bull Ring centre in figure 142?

Apart from these 'weather-free' centres, other parts of the city have been converted from congested streets into 'pedestrian areas'.

27 If you look carefully at figure 143, you can see the same features in both photographs to show that they are the same street.

28 Apart from the absence of vehicles, in what ways is the 'pedestrianised' street different from when it was open to traffic?

29 Do you think the street is more attractive 'before' or 'after'? What are the reasons for your opinion?

30 Vehicles are not entirely banned from the pedestrian streets. Besides buses, what sort of vehicles are likely to be allowed in 'for access'?

Even away from the pedestrian streets there are footways and under-

Figure 142 The Bull Ring
(a) (above) exterior
(b) (middle) interior
Figure 143 (right) Cherry Street, Birmingham (a) before (b) after

The Rape of Manchester

Some years ago I used to visit Birmingham frequently. Last March I went there again. I didn't recognise it – any part of it. So far as I could see there was nothing left at all of its sooty centre. It was as though, in the interval, someone had put a lid over the city, waved a wand, and then taken the lid off, revealing a totally new and utterly dull concrete relief of office blocks and roads.

Birmingham, however, is only the beginning of a huge and rapid commercial offensive that is about to wipe the nineteenth-century cities off the face of the North. New development will be of the most negligent and dreadful kind. Glasgow, Newcastle, Liverpool, Leeds, Manchester – Manchester most of all, according to the signs – are going to go the way of Birmingham.

What's happened to Manchester? says the man behind the desk in the planning office. What's happened? Why, it's going – going, going, *gone*. (Whatever he may say in public, this is how he talks in private.) Things have changed, you know. We aren't living in the nineteenth century now – you can't stop progress. One's got to keep up with the times. Land means money. Not just *money* – it's a *goldmine*. Yes, I know there are thousands of homeless but, be reasonable, how can you build houses in the centre when there's so much money lying about? The bigger the buildings, the bigger the returns – that's elementary arithmetic. Same with the shops: the little businesses are out, its multiples now – they can pay the rent. See that tower, boy, see the neon sign? One box on top of another box – just a pile of boxes, all the same. Don't try to be different. Keep to the straight and narrow. You can't go wrong. Stick to the rules. That's a nice clean building – that's *good* . . .

No one questions the principle that the local authorities – and their planning experts – are bound to make the best use they can of the property they control. But the 'best use' ought to include something besides immediate profit: it should include an architectural background that is good to live with. Decisions made by councillors should benefit society as a whole, not merely the developer. If, for instance, people were less densely packed in the city centre, heights of buildings and land values would also be reduced. The rest is then up to the architects. If the councillors used their considerable powers to induce developers to recognise that they, too, had a responsibility to the public – and it is the public that has to look at their buildings – they would have to take the business of finding much better architects very seriously; they would also have the reward of being really proud of their buildings. Time and money would be saved, endless public inquiries would be avoided.

What is happening to Manchester is a familiar story: it is happening to most of the industrial towns and cities, in some degree. To start with, there seem to be two Manchesters: a very rich one and a very poor one – not much in between. The rich revolve round the hotels, the poor are confined to their slums. The conflicts are not new: they are, on the whole, between people and cars, neighbourhoods and ringways, conservation and expansion. But since planning policy backs commercial and city council interests first, and housing last, Manchester's centre is set aside for tall office blocks and hotels – and land values are as high as the buildings.

The planners' disinclination to reverse this artificial situation is also familiar. And so are the consequences – a dead centre at night, 'nothing to do after ten', the poor on the periphery, the rich tucked away in villages anything from five to 15 miles away. This situation brings with it a predictable traffic problem: jams of cars converging on the city down one-way streets and trunk roads in the morning, and abandoning it (or attempting to) for the suburbs and the country in the evening. And then, of course, the traffic solutions follow well-used patterns: indestructible concrete motorways, elevated or in cuttings, tearing the old city's Victorian fabric to shreds. Such operations are usually described as improvements.

STEPHEN GARDINER

Figure 144

Figure 145 Pedestrian ways across a traffic island

passes. In many cases the centres of the big roundabouts have been made into attractive sunken gardens, or miniature shopping centres.

One would think that such a comprehensive scheme could not fail, and the City of Birmingham is proud of it. Yet it has its critics. Building these huge roads meant the demolition of buildings which many would have liked to preserve, such as housing and small-scale industry. Again, opponents feel that urban motorways simply encourage more vehicles to use them, so that eventually they too will become choked. This has happened in the USA so there seems no reason why it should not happen here.

Another problem is land prices If millions of pounds are spent on redevelopment schemes, anything that is built is going to cost a great deal. Some users of land simply cannot afford the rent. Even in the new shopping centres, the smaller shopkeepers complain that the rents are too high. Thus there is a great deal of economic pressure to build multi-storey office blocks and car parks, which can afford the costs of central locations.

One opinion is expressed clearly by the newspaper article in figure 144. Read it carefully.

31 This article is mainly about Manchester, but what are the author's opinions about Birmingham?

32 Are Birmingham and Manchester the only cities to worry the writer?

33 What do you think is meant by '. . . an architectural background that is good to live with'?

34 Name some of the features of cities which have been dealt with in this chapter and the previous one and which Stephen Gardiner writes about in this article.

Figure 146 The Old Wellington Inn and Sinclair's Oyster Rooms in central Manchester

Even in Manchester, despite Stephen Gardiner's fears, they do preserve *some* old buildings! (Figure 146.)

Nottingham's traffic problem

If Birmingham, Manchester and the rest do not have the answer, then who has? The building of indoor shopping centres and the paving of streets mainly for pedestrian use can hardly be criticised. But if a new road network is not to be built, what is to happen to all the traffic? Well . . . you discourage it from entering the town centre at all.

In Nottingham, the plan for a £100 000 000 new road system has been scrapped and replaced by an attempt to improve what is already there. Sub-standard housing needs a lot of money to make it habitable by today's standards, but then, £100 000 000 *is* a lot of money.

In Nottingham, ordinary traffic is being restricted in the city centre and a great deal of money is being spent making the bus service efficient and *free*. A double-decker bus carries as many people as would travel in fifty cars. Imagine the difference in traffic congestion, noise and air pollution if there were only buses. Yet because buses are often caught in rush hour traffic, they are usually slow, infrequent and poor value for money. Therefore, many travellers use their cars. This means more traffic jams and less money for the bus companies. The fares have to be increased, and, because of the traffic, the buses are slower. Therefore, more people use cars . . . and so on. It is a vicious circle.

To break this spiral, buses have to be given special treatment. Bus lanes have to be created, buses have to be given priority at traffic lights by special electronic equipment, and where buses do mix with other traffic there must be as little congestion as possible.

35 If you wanted to drive through central Nottingham, which route would you take, following the arrows?

Because this scheme does not involve massive rebuilding (although there are new shopping centres), it means that alterations can be made as time goes on and more knowledge is gained about how well it is working. A human **scale** to buildings and roads can be kept. Even in Nottingham, though, there are complaints. The rents in the shopping centres are too high. Other traders have suffered a drop in trade. They blame this on the reduced number of cars, the absence of on-street parking and the counter attractions of the new shopping centres, which are some way away from the traditional shopping streets.

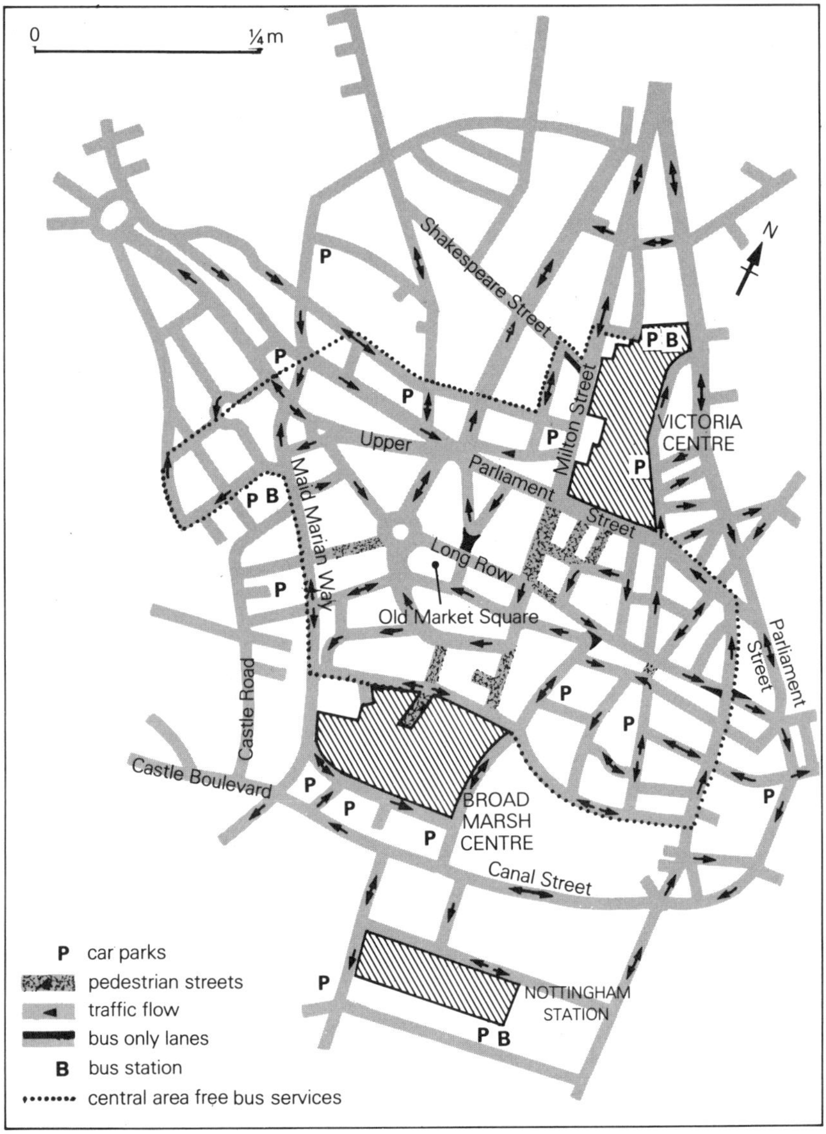

Figure 147 Nottingham's traffic scheme

36 Can you find the new shopping centres in figure 147?

Which system is the better, Nottingham or Birmingham? You are likely to find something similar to one of these in your nearest big city. It may already be in operation, although it is unlikely to be completed. Find out about it from the planning office. Then decide what *you* would do about the problem of traffic, poor housing and pollution.

PROJECTS

1 Conduct a survey of people who live in a newly-developed area. Keep your questions simple and, naturally, always be polite and explain what you are doing. Your

questions could be something like:

(a) Do you like living here?
(b) Name three things which you like about this development.
(c) Name three things which you dislike.
(d) Do you prefer this area to the one you came from?

In conducting the survey, make sure you find out the opinions of different types of residents. For example, make sure you interview people in different age groups and people in different types of dwelling (tower blocks, maisonettes, houses and so on).

2 The pictures in figure 148 are from an exhibition in the London borough of Camden. Using the information you have gained from this chapter and the previous one, explain in your own words the

Figure 148

problems which they illustrate and the cause of those problems.

3 Get hold of the *Bulletin of Environmental Education* (*BEE*) number 16–17, August–September 1972. This may be in your school library, or in your local library. If not, write for it to: The Education Section, Town and Country Planning Association, 17 Carlton House Terrace, London, SW1Y 5AS.

This issue of the magazine is concerned with making town trails. When you have read the articles, see if you can devise a town trail for your nearest town. Perhaps you can make more than one.

4 Have a class debate on the motion:

The best way to deal with the problems of a city centre is to knock it all down and start again.

9 I don't like it here

What is migration?

Have you ever moved house? If so, how far did you move? Was it within the same town? Perhaps it was from inside the town to a suburb or village further out, or you might have moved a long way, from one part of the country to another. It could be that you moved into the UK from some other country. Going to live somewhere else is called **migration**.

All the big conurbations of Britain show signs of people moving out from the centre to the surrounding countryside. This is partly caused by urban redevelopment, but many of the people move out of their own accord.

The reasons are many; poor housing, high rents or land prices, lack of space and high prices in the shops. People may feel that they prefer village life and the quiet of the rural landscape.

Migration in the South East

Figure 149 is a table from the census of 1971. The whole table has been included here so that you will learn how to sort out the parts you want from those you do not. This will enable you to visit your local library and use the census reports with ease.

1 First look at the total, at the top of the table, for the whole of Greater London. We are only interested in the columns headed 'Persons' (that is, not the separate males and females columns), for the years 1951, 1961 and 1971.

What was the total population of Greater London in each of the three years?

2 What, therefore, is happening to the population of Greater London?

3 Your answer to question 2 will show a contrast with the pre-war years. What was happening at that time? Look back to page 91 in the previous chapter if you need to be reminded.

4 Next, look down the list of London boroughs to see which have lost the most people between 1951 and 1971. The subtraction needed has been done for you and is shown in red on figure 149.

5 You will have seen that some London boroughs have actually gained population. Look at figure 150 to see whereabouts in London are the boroughs which lost most people and those which gained most.

6 You could now draw a map to show the change in London's population. Trace figure 150 and shade in red those places which have lost population. Shade in green those which have gained.

7 Although the map drawn for question 6 shows something of the pattern of population changes in London, it is perhaps too simple.

Another version could be drawn as follows:

green—population increase of 15 000 and over

yellow—population increase of less than 15 000 *and* population decrease of less than 15 000. That is, boroughs whose population has changed little over the 20 years.

brown—population decrease of 15 000 to 45 000

red—population decrease of over 45 000.

8 Using the map from question 7, where in London are most of the boroughs which have lost more than 45 000 people? Where are the

Figure 149 (opposite) Population 1951–1971 and intercensal variations

	Area	Population							Intercensal increase or decrease (-)					Net 1961 population of areas added or deducted (-) in the intercensal period 1961–1971
		1951	1961			1971			Amount	Per cent per year				
											1961–1971			
		Persons	Persons	Males	Females	Persons	Males	Females	1961–1971	1951–1961	Total	By births & deaths	Balance	
	a	b	c	d	e	f	g	h	j	k	l	m	n	o
	‡GREATER LONDON	8 196 807	7 992 443	3 810 952	4 181 491	7 452 346	3 578 238	3 874 108	−540 097	−0·25	−0·70	1·01	−1·32	New area
	City of London	5324	4767	2386	2381	4245	2191	2054	−522	−1·09	−1·15	−0·42	−0·70	
	London Boroughs													
−28 630	‡Barking	189 430	177 092	87 569	89 523	160 800	78 494	82 306	−16 292	−0·67	−0·96	0·30	−1·29	New area
−13 878	‡Barnet	320 438	318 373	148 209	170 164	306 560	144 799	161 761	−11 813	−0·06	−0·38	0·32	0·72	New area
+11 678	‡Bexley	205 400	209 893	102 039	107 854	217 076	105 846	111 230	7183	0·22	0·34	0·62	−0·30	New area
−30 424	‡Brent	311 081	295 899	142 652	153 247	280 657	136 490	144 167	−15 242	−0·50	−0·53	0·92	−1·59	New area
+37 605	‡Bromley	267 771	293 394	138 113	155 281	305 377	145 622	159 755	11 983	0·91	0·40	0·50	−0·10	New area
−51 581	‡Camden	258 318	245 707	113 693	132 014	206 737	96 542	110 195	−38 970	−0·50	−1·71	0·41	−2·21	New area
+23 437	‡Croydon	310 433	323 927	151 981	171 946	333 870	159 151	174 719	9943	0·42	0·30	0·54	−0·25	New area
−9582	‡Ealing	310 690	301 646	145 265	156 381	301 108	147 056	154 052	−538	−0·29	−0·02	0·66	−0·72	New area
−20 108	‡Enfield	288 112	273 857	130 235	143 622	268 004	128 243	139 761	−5853	−0·50	−0·22	0·33	−0·56	New area
−17 885	‡Greenwich	235 549	229 810	111 236	118 574	217 664	105 309	112 355	−12 146	−0·25	−0·54	0·36	−0·93	New area
−45 070	‡Hackney	265 349	257 522	124 120	133 402	220 279	105 764	114 515	−37 243	−0·30	−1·55	0·90	−2·69	New area
−54 236	‡Hammersmith	241 431	222 124	107 140	114 984	187 195	89 822	97 373	−34 929	−0·83	−1·70	0·60	−2·44	New area
−37 238	‡Haringey	277 316	259 156	122 569	136 587	240 078	115 059	125 019	−19 078	−0·67	−0·76	0·79	−1·68	New area
−16 279	‡Harrow	219 494	209 083	99 208	109 875	203 215	97 710	105 505	−5868	−0·48	−0·28	0·47	−0·78	New area
+55 602	‡Havering	192 094	245 598	120 023	125 575	247 696	121 629	126 067	2098	2·48	0·09	0·82	−0·79	New area
+24 576	‡Hillingdon	210 312	228 361	112 052	116 309	234 888	115 131	119 757	6527	0·82	0·28	0·68	−0·42	New area
−4119	‡Hounslow	211 075	208 893	100 875	108 018	206 956	101 100	105 856	−1937	−0·10	−0·09	0·56	−0·68	New area
−69 128	‡Islington	271 002	261 232	127 271	133 961	201 874	98 245	103 629	−59 358	−0·37	−2·54	1·01	−3·96	New area
−30 890	‡Kensington and Chelsea	219 117	218 528	96 490	122 038	188 227	85 488	102 739	−30 301	−0·03	−1·48	0·54	−2·13	New area
−6090	‡Kingston upon Thames	146 615	146 010	67 925	78 085	140 525	66 587	73 938	−5485	−0·04	−0·38	0·34	−0·74	New area
−39 448	‡Lambeth	346 964	341 624	163 605	178 019	307 516	147 985	159 531	−34 108	−0·15	−1·05	0·85	−2·06	New area
−34 597	‡Lewisham	303 071	290 582	138 736	151 846	268 474	128 132	140 342	−22 108	−0·42	−0·79	0·63	−1·51	New area
−22 816	‡Merton	200 140	189 013	88 819	100 194	177 324	84 205	93 119	−11 689	−0·57	−0·64	0·24	−0·90	New area
−56 627	‡Newham	294 017	265 388	130 503	134 885	237 390	116 800	120 590	−27 998	−1·02	−1·11	0·67	−1·90	New area
−17 013	‡Redbridge	256 902	250 080	118 668	131 412	239 889	115 218	124 671	−10 191	−0·27	−0·41	0·35	−0·79	New area
−13 472	‡Richmond upon Thames	188 100	180 949	84 228	96 721	174 628	82 244	92 384	−6321	−0·39	−0·35	0·23	−0·60	New area
−75 500	‡Southwark	337 638	313 413	151 364	162 049	262 138	125 520	136 618	−51 275	−0·74	−1·77	0·55	−2·46	New area
−6657	‡Sutton	176 151	169 095	78 868	90 227	169 494	79 857	89 637	399	−0·41	0·02	0·25	−0·23	New area
−65 014	‡Tower Hamlets	230 790	205 682	101 863	103 819	165 776	82 104	83 672	−39 906	−1·14	−2·13	0·49	−2·76	New area
−40 788	‡Waltham Forest	275 468	248 591	118 156	130 435	234 680	112 717	121 963	−13 911	−1·02	−0·57	0·37	−0·98	New area
−28 625	‡Wandsworth	330 883	335 451	158 906	176 545	302 258	144 134	158 124	−33 193	0·14	−1·04	0·60	−1·76	New area
−60 584	‡Westminster, City of	300 332	271 703	126 185	145 518	239 748	113 044	126 704	−31 955	−0·99	−1·24	0·21	−1·48	New area

Notes: (1) All figures relate to the areas as constituted in 1971. (2) Changes since 1961 are indicated by symbols – * boundary, † name ‡ newly constituted area.

HERTFORDSHIRE
ESSEX
BUCKINGHAM
KENT
SURREY
Enfield
Barnet
Harrow
Haringey
Waltham Forest
Redbridge
Havering
Hillingdon
Brent
Camden
Islington
Hackney
Barking
Newham
Tower Hamlets
City
Kensington and Chelsea
Westminster
Hammersmith
Ealing
Hounslow
Richmond-upon-Thames
Wandsworth
Lambeth
Southwark
Greenwich
Bexley
Lewisham
Merton
Kingston-upon-Thames
Sutton
Croydon
Bromley

Figure 150 (opposite) Greater London boroughs

boroughs which have gained more than 15 000?

We have found that Inner London is losing people and some outer London boroughs are gaining. In fact, many of the gains are because those boroughs have new housing estates in them to provide homes needed because of redevelopment in the inner areas.

But the gains in Outer London are not nearly enough to account for all the losses. Where have the rest of the people gone? We know from the previous chapter that many have gone to expanded towns or to new towns. A far greater number have moved to existing towns and villages outside the green belt. This outward, or 'centrifugal', movement from city centres is a common pattern of migration around all the big cities of the UK. There are, however, other migration patterns.

Movement between regions

Look at figure 151.

9 Which regions had a net gain in people through migration?
10 Which have had a net loss?

The word 'net' is used in these questions because all regions lose some people and all gain some. The net change is the difference between gains and losses.

11 Trace figure 152 and shade in red those regions which have had a net loss through migration, and shade in green those which have had a net gain.

Does your map show any similarity to figure 44 (page 39).

You should not expect the two maps to be exactly alike. Because the regions in figures 151 and 152 are so large, they disguise changes in population *inside* each region.

Figure 151 Population changes by migration, 1961–71

	Net annual migration thousands
North	−10·8
Tyneside conurbation	− 8·1
Remainder of North	− 2·7
Yorkshire and Humberside	− 7·0
West Yorkshire conurbation	− 3·7
Remainder of Yorkshire and Humberside	− 3·3
East Midlands	+ 7·1
East Anglia	+12·1
South East	− 3·7
Greater London Council area	−98·2
Outer Metropolitan area	+43·8
Remainder of South East	+50·7
South West	+22·4
West Midlands	− 1·3
Conurbation	−17·8
Remainder of West Midlands	+16·5
North West	−11·4
Merseyside conurbation	−20·3
South East Lancashire conurbation	−13·3
Remainder of North West	+22·2
Wales	− 0·4
South East Wales	− 4·0
Remainder of Wales	+ 3·6
England and Wales	+ 7·1
Scotland	−32·5
Central Clydeside conurbation	−25·8
Remainder of Scotland	− 6·7
Northern Ireland	− 7·1

For the moment, however, let us think about the migration pattern shown by your map. Most of the areas that formed the outer region in figure 44 have lost people through migration. Those forming the centre have gained people. Or have they? The South East and the West Midlands, the two most prosperous areas, show a net loss. At this point, you have to remember centrifugal migration. Look at the migration figures for the South East apart from Greater London and from the West Midlands apart from the conurbation.

12 Which two regions are now attracting most migrants?

At this point you may recall that part, at least, of these two regions were in the 'outer' zone of Britain. How, then, can they be attracting immigrants? The answer was mentioned earlier. The regions are too big to give an accurate picture. Inside any one region the prosperous parts attract people. They are usually around the main towns. But even in growing regions the **remote** parts are losing population.

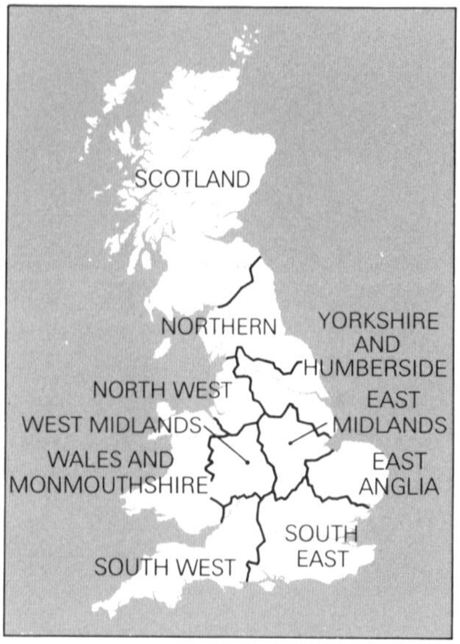

Figure 152 Standard regions of Britain, 1971

Areas losing people

Figure 153 is a map produced by the Department of Economic Affairs in 1969 to show areas where the constant loss of people has caused many problems.

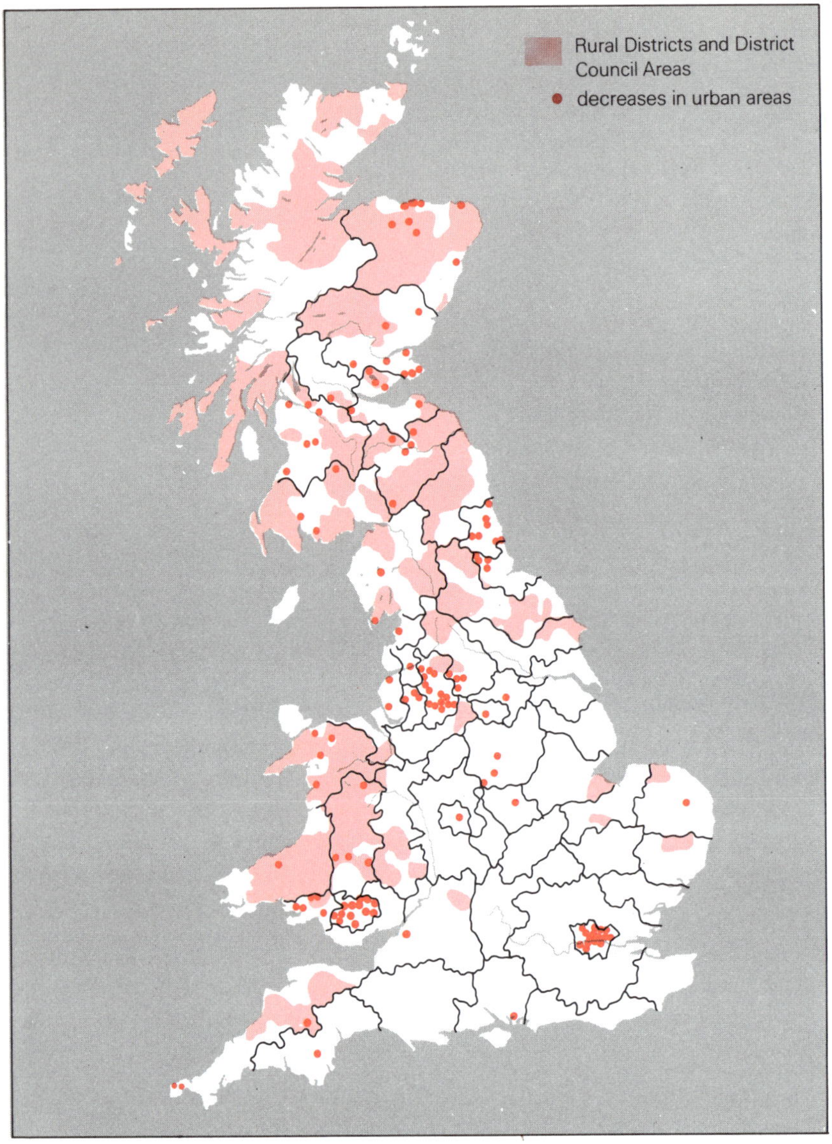

*Figure 153 **Persistent population decrease, 1951–61, 1961–66***

13 List some of the reasons why the places marked with dots have been steadily losing population.
14 Look back to figure 151. Which region had the biggest loss of people through migration?

This region is so large that the figures in the table do not tell the full story. As you can see from figure 153, not all of Scotland shows a persistent decline.

15 Look again at figure 151. Which part of Scotland has most people leaving it?

This movement is centrifugal migration, similar to that in the other big conurbations of the UK. It is worse than many because Glasgow is one of the cities based on industries which are now much less prosperous than they were. Also, the Clydeside conurbation is further from the main markets in the central area of the UK.

If Glasgow is the main place which is losing population, what of the rest of Scotland? It, too, is losing people by migration, but at a much slower rate.

The Scottish Highlands – rural depopulation

There is one part of Scotland which has been losing people, both by migration and by having more deaths than births, for well over a century. This is the Highlands.

16 Do you like the view in figure 154? If you have not been there, would you like to go to the area for a holiday?
17 What do you know of the Highlands at the moment? Write down a few sentences to describe the area under the following headings:

Figure 154

a) scenery; b) history; c) way of life of the people; d) dress.

Not only have the Highlands been losing people for a long time, but there are not many people to lose. Scotland has a total population of 5 217 400 and of these, almost 4 000 000 live in the central Lowlands. In the area officially called the Highlands (figure 155(a)) there are only 279 600 people.

Even though the official area is smaller than the physical one (figure 155(b)), it is still very large. In fact it occupies 47 per cent of the total area of Scotland. Since it includes the Hebrides, the Orkneys and the Shetlands, parts of the Highlands are a very long way from the centres of population and one village is often a long way from the next.

18 Look up in an atlas the Hebrides, the Orkneys and the Shetlands.

19 How far is it from Stornoway, from Kirkwall and from Lerwick to (a) Inverness, (b) Glasgow, (c) London?

20 Look up in an atlas the town and the islands named in the timetable shown in figure 156. What would it be like to live on one of these islands?

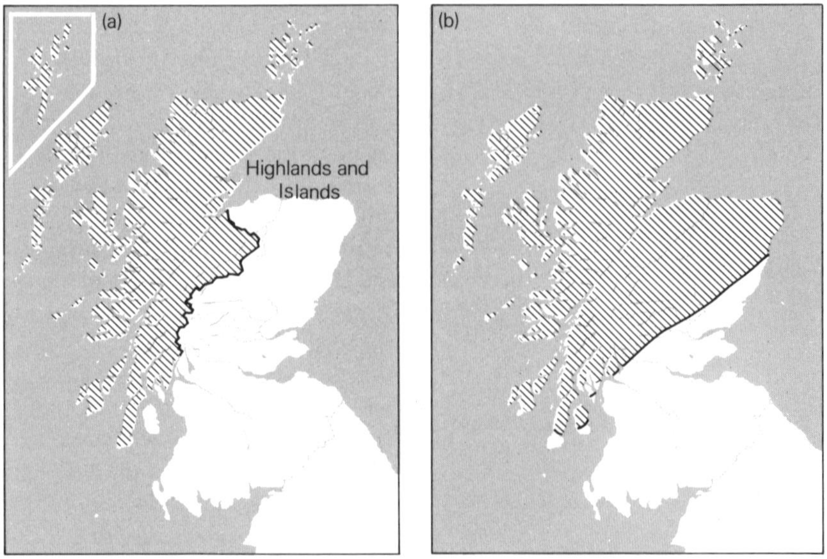

Figure 155 (a) The Highlands: the 'official' area
(b) The Highlands: the physical region

21 Do any of the islands have a boat every day of the week?

22 How frequently do you think it is possible for the islanders to receive letters?

23 This is the summer timetable. Do you think there will be more or fewer boats in winter?

If you think that these islands are fairly remote from Mallaig, you must now consider that Mallaig itself is over six hours by train away from Glasgow. That is longer than it takes to get from Glasgow to London!

You also have to consider that these are the *Inner* Hebrides. The outer islands are even more remote, even though some of them are served daily by air. Similarly, the Shetlands and the smaller islands of the Orkneys are a very long way even from the mainland of the Highlands.

Apart from this physical remoteness, there are not many people in the islands. The populations of those in the timetable are:

Eigg:	69	Rhum:	40
Muck:	24	Canna:	22

This small number of people cannot

Figure 156 Mallaig/Small Isles services

CALEDONIAN MACBRAYNE LTD. — Car and Passenger Services 3 May until 23 October, 1976

SKYE AND OUTER ISLANDS LEAFLET

Alteration to Mallaig — Eigg — Muck — Rhum — Canna service

The service between Mallaig—Eigg—Muck—Rhum and Canna as detailed in the above publication will cease after Monday, 31 May, 1976

THE UNDERNOTED SERVICE WILL OPERATE FROM 1 JUNE UNTIL 23 OCTOBER, 1976

	MONDAY		TUESDAY	WEDNESDAY	THURSDAY		FRIDAY		SATURDAY	
MALLAIG	—	1230	1230	1230	0600	1230	—	1230	0500	1300
ARMADALE	—	1310*	1310*	1310*	—	1310*	—	1310*	—	1340*
KYLE OF LOCHALSH	—	—	1510	—	—	1510	—	1510	—	—
EIGG	—	1440	—	1440	—	—	—	—	—	1510
MUCK	—	—	—	1525	—	—	—	—	—	†
RHUM	—	1610	—	1640	—	—	—	—	—	1625
CANNA	—	1710	—	1740	0830	—	—	—	0730	1740
RHUM	—	—	—	—	0930	—	—	—	0830	—
MUCK	—	—	—	—	—	—	—	—	0945	—
EIGG	—	—	—	—	1045	—	—	—	1030	—
KYLE OF LOCHALSH	0930	—	1530	—	—	—	0930	1530	—	—
ARMADALE	—	—	—	—	—	—	—	—	—	—
MALLAIG	1130	1940	1730	2010	1215	—	1130	1730	1200	—

CODE: * Connects with 1400 Bus ex Armadale † Private operator connects from Eigg

CALEDONIAN MACBRAYNE LTD., THE PIER, GOUROCK PA19 1OP

support many services. Look back to chapters 6 and 7 to remind yourself about the number of people needed to keep different types of shops going.

24 Think back over the last week and make a list of the things you did for recreation.

25 How many of the items on your list could be done on an island with a total population of 40?

26 Are there any things you would like to do which could only be done in a remote place?

It is the remoteness and absence of services which is perhaps the main cause of people leaving. There are very few senior secondary schools in the remoter parts of the Highlands, even on the mainland. Anyone in the 14 to 19 age-group who wants education often has to leave home during the week. There are very few centres anywhere in the Highlands for professional training, or technical and apprenticeship schemes. There are no universities or colleges of education in the Highlands at all. It is quite likely that you have thought about what you will do when you leave school. Suppose you lived on the island of Barra. If you wanted to go to technical school, you would probably have to go to Fort William or Oban. If, then, you wished to take up a craft apprenticeship, you would have to go to one of the larger Highland centres, like Fort William or Inverness. All of these options would involve leaving home. Then, if you became a skilled craftsman, or were trained for a profession, what would you do with your skills?

Another reason why people leave the Highlands is that there is not much work apart from agriculture. The type of agriculture in much of the Highlands is organised in an inefficient manner called **crofting**. Crofts are often farmed in very small strips of land which are owned by several people or families.

In many cases the croft is not large enough to occupy one man full-time So what else can he do? Many crofting lands are on the coast, so he can fish. But to make a good living from fishing these days, you need a modern boat and a lot of expensive equipment. For this reason, most of the UK's fishing is now organised in large fleets, based on ports outside the Highlands.

It is not really surprising, therefore, that young people leave. And if the young leave, there will not be so many left who are the right age to have babies. When the old people die, there is no one to take their place. There are more deaths than births. This loss of people in remote areas is called **rural depopulation**.

Plans for the Highlands

27 Before reading on, see if you have any ideas about what could be done to help the Highlands. Imagine you have to present a report to the government, stating your plans for the Highlands and giving reasons for the suggestions.

Now let us look at what is being done about the rather gloomy picture of the Highlands which we have so far seen:

(a) **Farming**: an organisation called the Crofters' Commission is trying to make crofting more profitable. Spectacular achievements have been made by the Highlands and Islands Development Board. This is an official body, set up in 1965, with wide powers.

(b) **Fishing**: by providing money to help with building boats and buying equipment, the Board has helped in a small revival of the fishing industry. This is important, because as well as work for the fishermen, it is also estimated that for every one fisherman, two jobs are created ashore.

28 What sort of shore jobs are connected with the fishing industry?

29 Look at figure 157. Which are the two chief activities that the HIDB has promoted, both in numbers of projects undertaken and in numbers of jobs created?

(c) **Tourism**:

30 In the centre of figure 158 is a group of large new buildings. What could they be for? Could a group of hotels manage to attract enough

Figure 157 HIDB (a) Number of projects approved 1965–72 (b) Jobs expected 1965–72

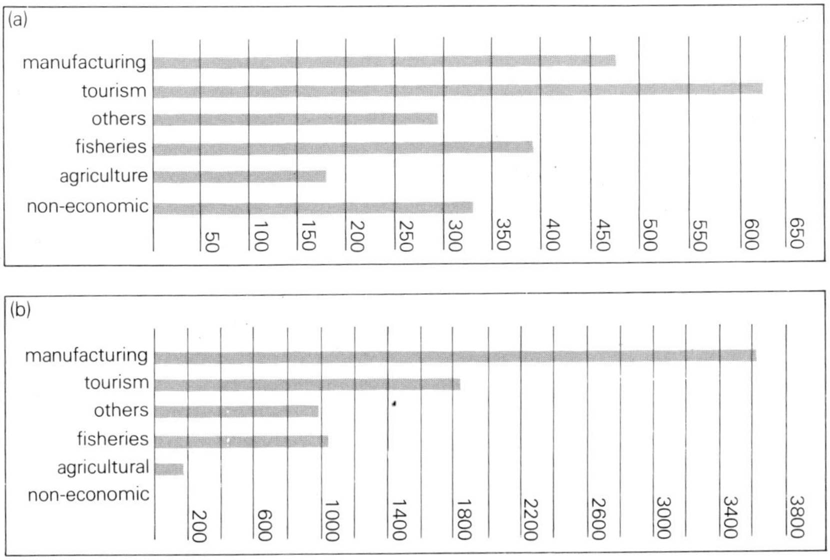

Figure 158

people in this wild landscape?

31 Can you think of any way in which one or more large hotels could attract sufficient people?

In fact the picture shows a 'tourist centre' in the 'village' of Aviemore. Sporting activities are provided all year round, though there is a strong accent on winter sports. There are hotels, shops, a theatre, ballroom, ice rink and swimming pool, all set in a 28 000 hectare National Park. It has become a winter and summer recreational centre and attracts conferences at off-peak periods.

32 Do you think that people would go to the Scottish Highlands in the middle of winter to go to the theatre?

33 Why might people go to an area such as this in winter?

Apart from major schemes, tourism provides jobs in services such as road improvements and shops. It also gives part-time income for crofters, farmers and others, who can rent their land for caravans or provide bed and breakfast.

(d) **Industry**: most of the effort in the area is going into providing work in manufacturing. In many cases, the industries which the HIDB supports are small, employing only a few people. They are very varied, ranging from help with Harris Tweed making on the Isle of Lewis/Harris, to a plastics factory at Evanton, or an ink factory at Campbeltown.

34 See if you can find these places in an atlas or AA book.

Although small, these industries can save a small township from decay.

The Board is trying to attract larger industries to three main growth areas.

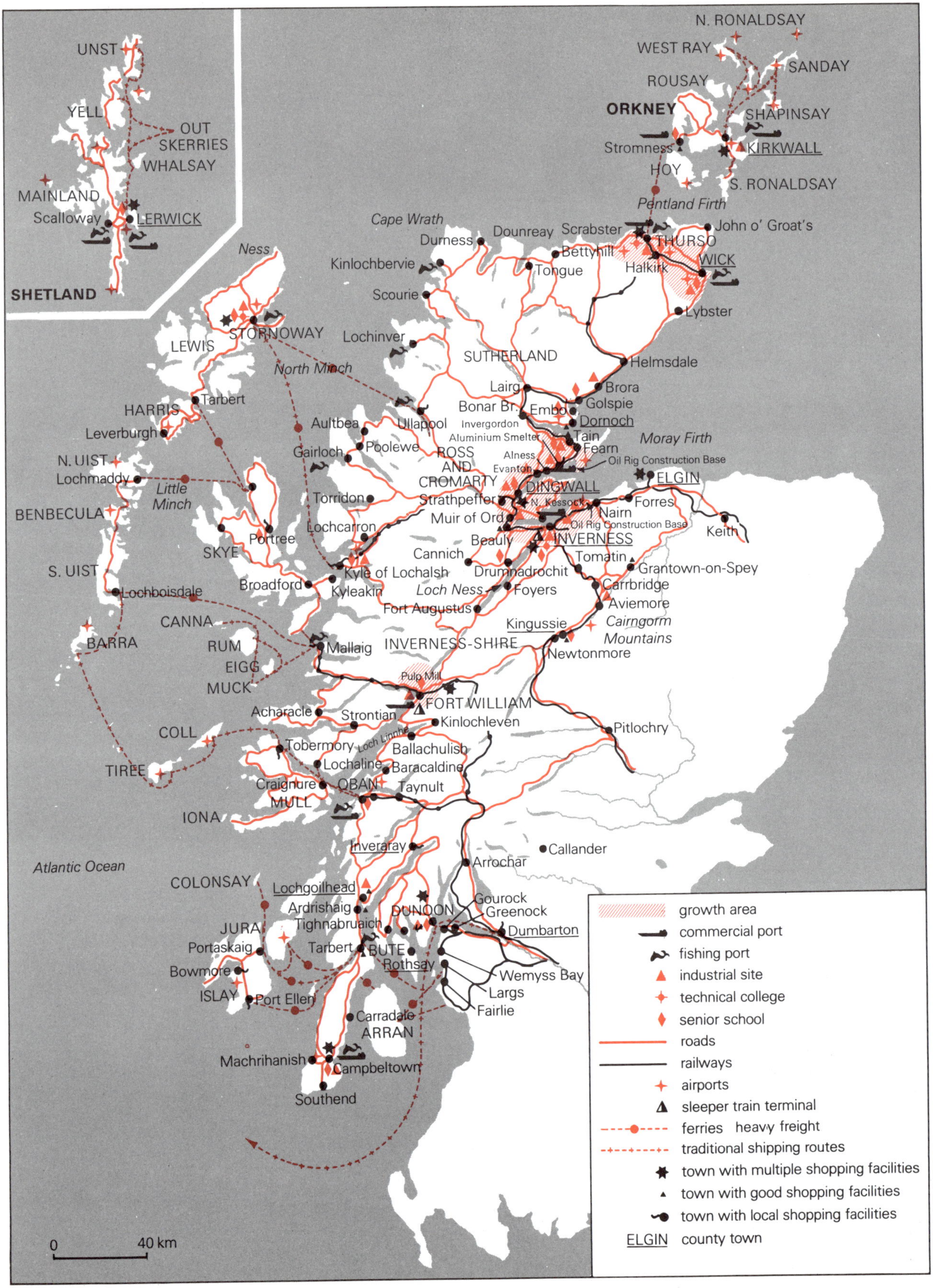

UNST
YELL
OUT SKERRIES
WHALSAY
MAINLAND
Scalloway
LERWICK
SHETLAND
N. RONALDSAY
WEST RAY
SANDAY
ROUSAY
ORKNEY
SHAPINSAY
Stromness
KIRKWALL
HOY
S. RONALDSAY
Pentland Firth
Cape Wrath
Durness
Dounreay
Scrabster
John o' Groat's
THURSO
Bettyhill
Tongue
Halkirk
WICK
Kinlochbervie
Scourie
Lybster
Ness
STORNOWAY
LEWIS
Lochinver
SUTHERLAND
North Minch
Helmsdale
Lairg
Brora
Golspie
HARRIS
Tarbert
Bonar Br.
Embo
Ullapool
Dornoch
Leverburgh
Aultbea
Invergordon
Aluminium Smelter
Tain
Moray Firth
Poolewe
ROSS AND CROMARTY
Fearn
Gairloch
Alness
Oil Rig Construction Base
N. UIST
Evanton
Lochmaddy
ELGIN
Little Minch
DINGWALL
Torridon
Strathpeffer
N. Kessock
Forres
BENBECULA
Muir of Ord
Nairn
Oil Rig Construction Base
Lochcarron
INVERNESS
Keith
Portree
Beauly
SKYE
Cannich
Tomatin
S. UIST
Kyle of Lochalsh
Drumnadrochit
Grantown-on-Spey
Broadford
Kyleakin
Loch Ness
Foyers
Carrbridge
Lochboisdale
Aviemore
Fort Augustus
CANNA
Kingussie
Cairngorm Mountains
RUM
BARRA
Mallaig
INVERNESS-SHIRE
Newtonmore
EIGG
MUCK
Pulp Mill
FORT WILLIAM
Acharacle
Strontian
Kinlochleven
COLL
Pitlochry
Tobermory
Loch Linnhe
Ballachulish
Lochaline
Baracaldine
TIREE
Craignure
OBAN
Taynult
MULL
IONA
Inveraray
Callander
Atlantic Ocean
Arrochar
COLONSAY
Lochgoilhead
Gourock
Ardrishaig
DUNOON
Greenock
JURA
Tighnabruaich
Dumbarton
Portaskaig
Tarbert
BUTE
Rothsay
Bowmore
Wemyss Bay
ISLAY
Largs
Port Ellen
Fairlie
Carradale
ARRAN
Machrihanish
Campbeltown
Southend
0
40 km
growth area
commercial port
fishing port
industrial site
technical college
senior school
roads
railways
airports
sleeper train terminal
ferries heavy freight
traditional shipping routes
town with multiple shopping facilities
town with good shopping facilities
town with local shopping facilities
ELGIN county town

Figure 159 (page opposite) HIDB map of the Highlands

35 Find these areas on figure 159.

The area around Fort William had some of the first industries in the Highlands, particularly whisky distilleries and a large aluminium works. The aluminium works was located there so as to be near deep water to import the ore and to have cheap supplies of electricity. At one time it was thought that because of the high rainfall, numerous lakes and high mountains a vast amount of electricity could be produced by water power, but more recently, it has been discovered that this is not the case. It is estimated that the total possible hydro-electric output would be only about half that of a single big power station in England! However, it is important that electricity is available in the Highlands so that even remote settlements are not without this one luxury.

(e) **Forestry**: until recently, the factory in figure 160 was perhaps the most spectacular development in the Highlands.

36 Can you think of any reasons why it should be located here? A clue to one of the reasons is the sub-heading of this section; another reason is suggested in the photograph, just behind the factory.

37 The mountain in the background is the highest in the UK. Do you know its name?

Timber producing is another activity which some parts of the Highlands are suited to and which helps to provide employment. Unfortunately, much of the land which is suitable for timber is also good for farming. As with so many things, a choice has to be made over the use of the land. Nevertheless the Forestry Commission has planted many hectares of trees where there was not too much conflict with farming.

Figure 160 Wiggins Teape pulp and paper mill at Fort William

Tree planting is one more way in which the drift away from the Highlands has been slowed down.

The Moray Firth area

The largest of the three growth areas is around the Moray Firth, at the north-east end of the Great Glen. There is no other focus of population in the Highlands of anything like the same size. This area has most of the largest towns in the Highlands and it is also a natural focus of routeways. It is not surprising, therefore, that the HIDB chose to encourage large industries to develop in the area, as well as many small ones.

The most suitable place for large industries is around the shores of Nigg Bay on the north side of the Cromarty Firth. Here there is deep water together with flat land, which is rare. In this area there is a huge aluminium smelter at Invergordon.

38 Find Invergordon and Nigg Bay on figure 159.

39 Why should an aluminium smelter need deep water nearby?

A large petrochemical works is also being built, and there are many smaller industries in the area. Because of this growth in manufacturing, the HIDB expects a great increase in the number of jobs, and a population growth from 70 000 to 300 000. With such an increase, the drift away from the Highlands would be reversed.

It is necessary, therefore, for you to understand that the decision to locate the factories in this area was a political decision taken by the HIDB with the support of the government, in order to provide work in the area. The physical geography alone would not have been enough to attract the industries.

Figure 161 *Work in progress on the dock for building oil platforms*

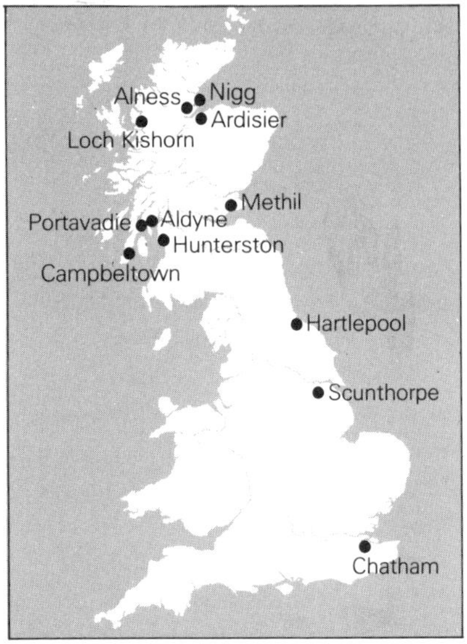

Figure 162 *Map of oil platform construction sites*

All this was before the most recent boost to Scottish jobs, already mentioned in chapter 3. In figure 161 you can see one of the results of this boost.

North Sea oil again

The excavation in figure 161 was once known as the 'largest hole on earth'. Another picture of it appeared in figure 38. Works such as these are called 'oil-related industries'; the main one is the building of oil-drilling rigs, which are enormous, and of production platforms, which are even bigger. To build these huge structures you need all the equipment of a shipyard; steel, power, deep water and a great many men. It also helps if you are reasonably near the North Sea oil finds so that you do not have to tow the rig too far.

40 How many of the oil platform construction sites shown in figure 162 are in the Highlands?

41 Is all the employment involved in building oil rigs likely to solve the problems of Highland depopulation?

Since 1966, the Highlands actually gained population for the first time, so it may seem that the answer to question 41 is 'Yes'. Yet it is not quite so simple. The gain in population has been in the development areas around Fort William and the Moray Firth. This may have made depopulation *worse* in the rural areas.

42 Can you think of any reasons why this should be so?

Not only is rural depopulation worse, but there is a shortage of skilled labour outside the oil-rig yards. It is said that in the Inverness area, people have to send 40 miles (65 kilometres) for a plumber!

Alas, this is not all. These platforms, as big as St Paul's Cathedral, need really deep water to test them and tow them out to sea.

43 Where, around the coast of the Highlands, are the most numerous stretches of sheltered, very deep water to be found?

When the Highlands were uplifted to their present height, they were tilted. Thus, the west is, in general, higher than the east. When the glaciers of the Ice Age cut down into the Highlands, they carved the deepest valleys on the western side. Then, later, when the sea level rose, many of these valleys were drowned to form deep sea lochs. Yet the valleys were long, often with sharp bends (caused by the glaciers following fault lines). (See chapter 13 in *Spotlight on Europe* in this series.) For this reason they are often very sheltered and so there was a proposal to build platforms at Drumbuie, near Kyle of Lochalsh.

44 In figure 163, how deep is the water in Loch Carron?

45 Is Loch Carron a sheltered site? (Do you know from which direction the prevailing wind blows in the UK?)

Figure 163 The Drumbuie area

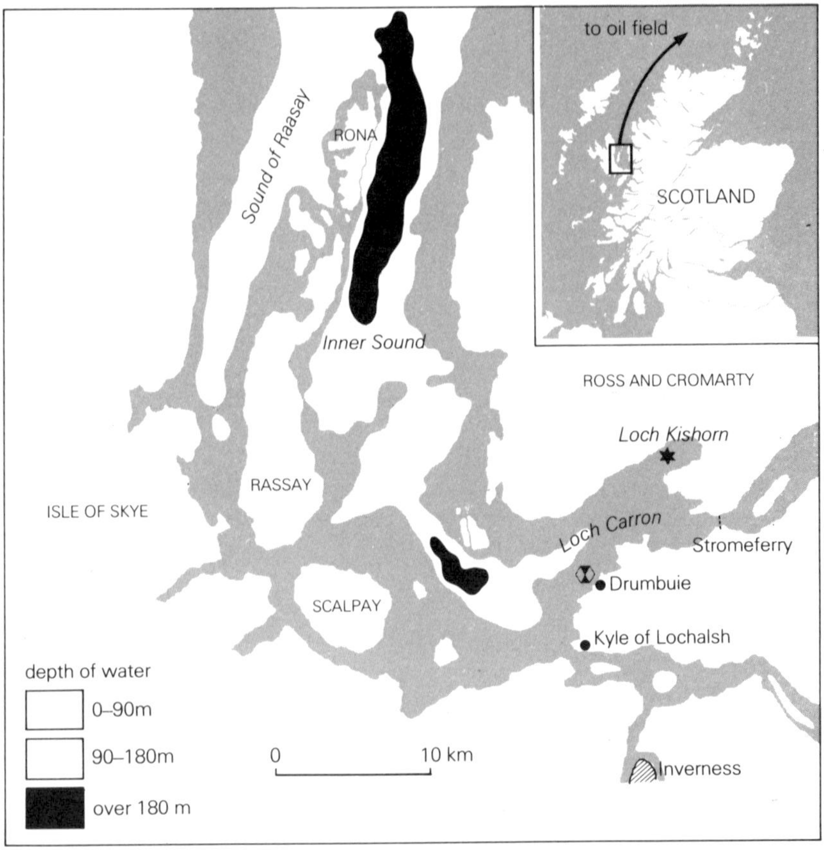

46 What do you think the scenery would be like in the Drumbuie area?

These same features of physical geography are some of the grandest in Britain. The scenery and remoteness are what attract the tourists. Visitors enable the local people to live in their traditional style with a reasonable income. To build the platforms would need a new town of 500 people, or so. Quite apart from the effects of the rig building on the scenery, imagine the effect of 500 newcomers on the way of life of the area. In the whole district there are only 1 200 people at present. In Drumbuie there are 25! And as the National Trust for Scotland says:

> Our civilisation needs such places and the people who live there.
> Let it be . . . *Fàg mar a tha e.*

After a lengthy inquiry the site was turned down by the minister. The opposition was successful . . . But, if you remember figure 162, you may be thinking that the Loch Kishorn rig building site was somewhere near Drumbuie.

The photograph in figure 164 looks much like any other Highland tourist photo, such as the one in figure 154. But the flat land on the far side of the loch is the site for the new platform building dock. The loch is Loch Kishorn, and if you wondered what the star was on figure 163, it was that piece of land.

The new site is a compromise. The whole nation needs the oil urgently. Even some of the local people would prefer 'development' to 'stagnation'. There is no simple answer.

Migration into the UK

Look at the names on the shops in figure 165. What is showing at the cinema? The picture would remind

Figure 164

Figure 165

you that there has been migration into the UK. Of these migrants, the ones who have had the most publicity are those whose culture and skin colour are different from the traditional British.

47 Do you know what is meant by the word 'culture'? If not, look it up in a dictionary.

These same people have been the subject of attack and prejudice. One of the reasons why some people are suspicious of the newcomers is that they tend to live in fairly small areas at high densities of population. Within these areas, they often retain their own culture and do not 'integrate' with the rest of the people.

48 Do you know what is meant by the word 'integrate'? If not, try to find out.

This is not the only reason for racial tension. There are many reasons, but *where* people live is a *geographical* reason and will therefore be looked at more closely.

Not all immigrants belong to the group shown in figure 165. The table in figure 166 shows the number of people who live in Great Britain but were born in the country shown.

Figure 166 The numbers, in thousands, are those people born in other countries who were living in Great Britain at the time of the 1971 census.

Foreign countries	980
'Old Commonwealth' (Canada, Australia, New Zealand)	143
India, Pakistan, Bangladesh	462
West Indies	304
Rest of 'New Commonwealth'	385
Irish Republic	709
Total	2983

49 Find out which countries are included in 'Rest of New Commonwealth'.

We saw in chapter 7 that people of the same culture like to live in the same area. If you add to this the fact that a particular culture is different from that of most other people in Britain, you would expect the people to group together even more closely. Often these are poor people who have come to look for better opportunities and there are very few areas where they can *afford* to live. Finally, you need to think about how immigrants make their decision on where they will live. You will remember (from chapter 1) that people decide to do things on the basis of what they *think* is true, rather than on what they *know*. In the light of all this, imagine you are a man or woman living at a great distance from the UK, who has decided to go and live there. What sort of place would you choose to live in? It is likely that you would be guided by others who had gone before you. It is also probable that when you moved you would first stay with friends or relatives who were already there. Then, when you found a house (or, more likely, a room) of your own, it would not be far from people you knew.

In the United States, areas where there are large numbers of people belonging to the same cultural type are called ghettos. This name originated in Europe where it referred to large concentrations of Jewish people. It is now wrongly applied to parts of the UK. The total numbers of immigrants are not large enough, nor is the degree of concentration great enough for the name to be used.

50 What is meant by 'degree of concentration'? If you know what it is, you may wonder how it is measured.

Figure 167(a) shows a map of the electoral wards of Bradford. Electoral wards are areas which are used when collecting information for the census,

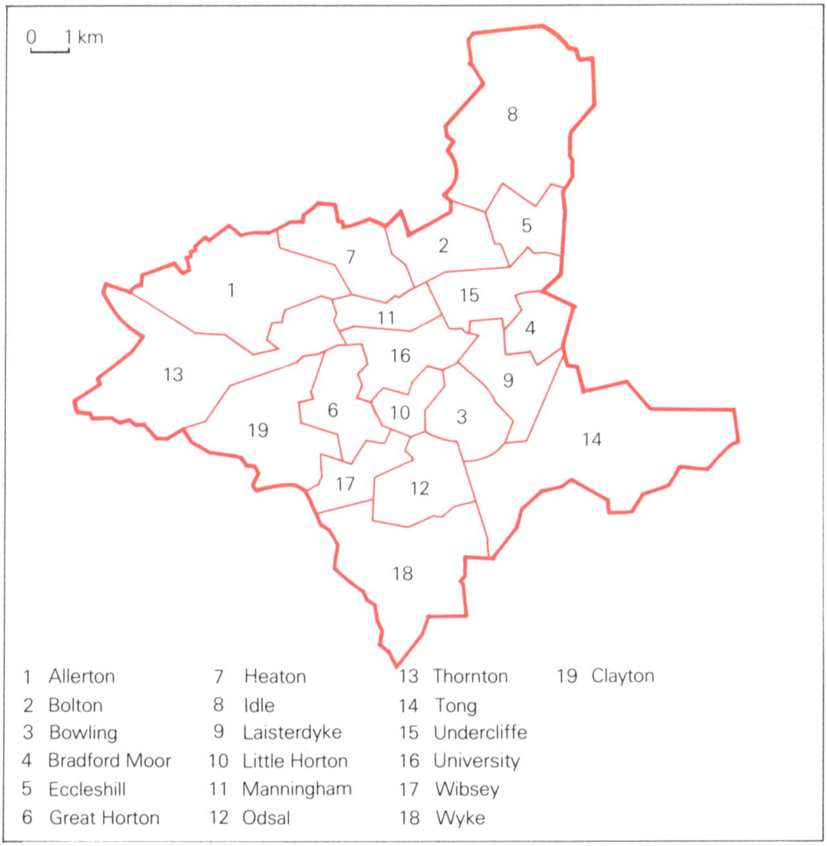

Figure 167 (a) (left) Bradford: electoral wards (b) (below) Numbers and percentages of residents whose parents were born in the New Commonwealth, 1971 census

Ward	Born in New Commonwealth		% of the total population whose parents were born in the New Commonwealth	Total population
	Males	Females		
Allerton	53	61	0·6	18230
Bolton	213	113	2·9	14753
Bowling	450	276	7·6	12518
Bradford Moor	1252	748	17·1	16044
Clayton	43	31	0·5	14337
Eccles Hill	58	43	0·6	15630
Great Horton	69	66	1·1	13944
Heaton	926	510	11·9	15606
Idle	46	32	0·4	15688
Laisterdyke	1061	576	17·0	11668
Little Horton	1476	867	18·4	17238
Manningham	4071	1757	37·4	19807
Odsall	46	32	0·6	14753
Thornton	30	34	0·5	13357
Tong	94	87	1·2	18889
Undercliffe	268	167	4·6	12987
University	3323	1722	40·1	16565
Wibsey	57	54	0·7	15256
Wyke	38	25	0·3	16702
TOTAL *for Bradford*	13574	7201	—	294175

as well as for electing local councillors. This means they are useful for working with figures from the census reports.

Below the map is a table which shows the numbers of people who were born in the New Commonwealth. Then there is another column showing the percentage of the total population whose *parents* were born in the New Commonwealth. This percentage should include immigrants (because their parents were probably born in the country they migrated from), and the children of immigrants.

Trace the map and then draw on your tracing a number of dots. Make each dot stand for fifty people and put the dots in the appropriate ward. In this way you will have drawn a dot distribution map which will show where the immigrant communities are concentrated.

In the case of Bradford, most New Commonwealth immigrants are Asians, from the Indian subcontinent. They are mostly concentrated in the older working-class areas, which are mixed in with the woollen mills. Bradford is part of the West Yorkshire conurbation which has been the chief wool manufacturing region of the UK for a long time.

Now look at figure 168 where two of the same sort of maps have been drawn for the London Borough of Ealing.

51 In figure 168 (a) which named area has the greatest concentration of immigrants?

52 Which named area had the greatest concentration in 1966?

53 Were there more or less immigrants in 1966 in the area you have named?

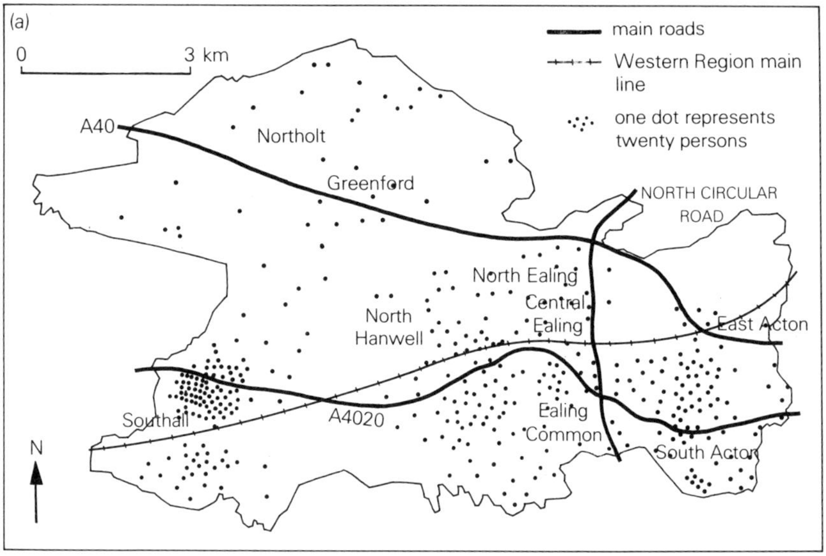

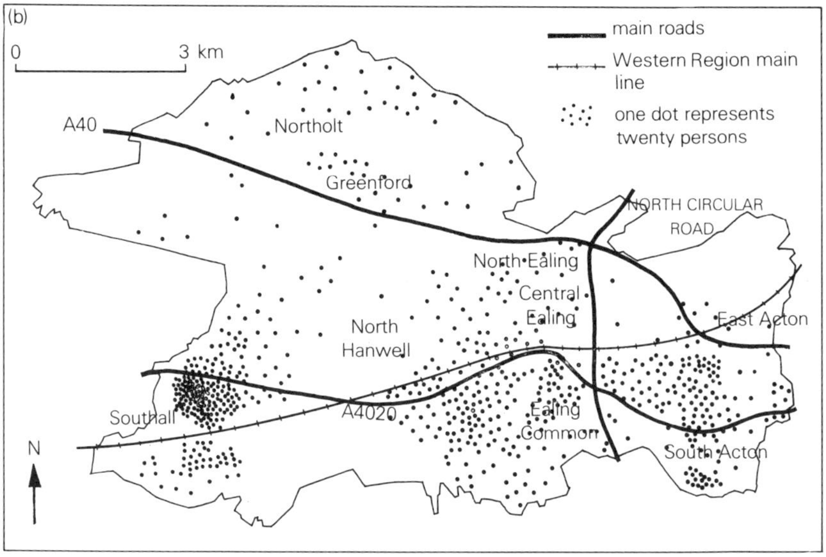

Figure 168 (a) London Borough of Ealing: distribution of New Commonwealth immigrants, 1961
(b) London Borough of Ealing: distribution of New Commonwealth immigrants, 1966

Your answers to these questions should confirm that new immigrants go to places where they know someone.

Perhaps these maps do not give a true picture. Knowing total numbers is not enough. This is where you can use the column of percentage figures in figure 167. In a ghetto (see page 114), the cultural group forms a very high percentage of the total population, say, over 80 per cent.

54 Which ward in Bradford has the highest percentage? Would you say that the ward formed a ghetto?

55 A good way to show the concentration of immigrants on a map is called the 'location quotient'. This is a way of comparing the percentage in each ward with the percentage for the town as a whole. For each ward, divide its percentage by the percentage for the whole town. If the ward's percentage is the same as the town's, the answer will be one. If the ward has less than the average, the answer will be less than one (a decimal, such as 0·27). If the location quotient is more than one, then that ward has an above-average concentration. Draw a Location Quotient (LQ) map for Bradford. Work out the LQ for each ward, then shade in red those wards which have an LQ of 2·5 and over. Shade in grey those between 1·5 and 2·49; use lines for those between 0·5 and 1·49 and use dots for those less than 0·5.

Figure 169 shows a similar map for the London Borough of Ealing.

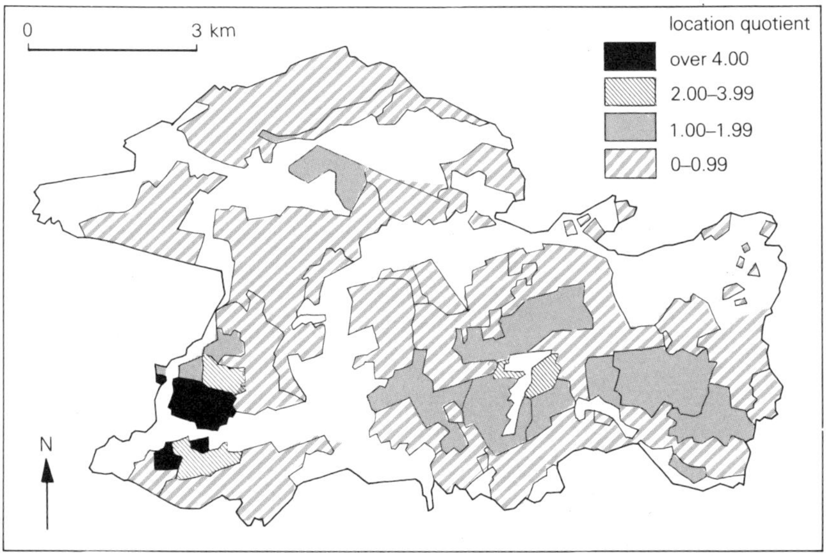

Figure 169 London Borough of Ealing: concentration of New Commonwealth immigrants, 1966

The empty spaces on the map are non-residential areas.

We now know that immigrants tend to group together in fairly small areas. But where are these areas? Immigrants in general, but New Commonwealth immigrants in particular, are mostly poor. Many wish to save their money, either to send back home to relatives, or to start their own businesses. For these reasons, the newly-arrived are often willing to accept poorly paid work and to work shifts. Most such jobs are to be found in the big conurbations and that is where most of the new immigrants go. London is the greatest attraction. In 1971 just over one million people living in Great Britain were those born in the New Commonwealth. Of these, nearly half lived in Greater London. About one third of the total lived in the West Midlands, the West Yorkshire and the South East Lancashire conurbations.

Within the conurbations, the poorest housing is often found in a 'collar' round the city centre (as we saw in Manchester, page 89). It is in these parts of the cities that we find the immigrants. Of course, there are exceptions, such as Ealing, but on the whole it is in these inner-city zones, amid dereliction and redevelopment schemes, that the newest immigrants live.

But are the immigrants destined to live in these places fore ever? Although it may seem like it to some of them, there is evidence that after a time the more successful move outwards, leaving room for new migrants.

It is likely that they do not move far, nor in any great numbers at any one time. Nonetheless, there does seem to be a process of **diffusion** at work, allowing a minority group to move to better housing. And why has this better housing become available? Perhaps the people from these houses are those we met at the beginning of this chapter, those who are moving out to the commuter towns, who are themselves looking for better housing.

In this section, immigrants have been referred to as if they were all the same. This illustrates one more problem. Each immigrant community has its own appearance, traditions and skills. Each group tends to live in its own area. The concentrations which have been dealt with are not 'immigrant' areas, but areas of perhaps Indian, Pakistani, Cypriot or West Indian culture.

There are differences between people; it is no use pretending there are not. But the differences of each group from white English people, and from each other, are very valuable. They need to be preserved and used for the development of society as a whole. And remember, 'society' is not all the same. It is now, and has long been, culturally, racially and socially mixed. This mixture can be both a strength and a weakness. It is up to you to see that it is the strength that is encouraged.

PROJECTS

1 Conduct a survey in your school to find out how many pupils have moved house in the previous twelve months.

Try to find out why their families moved. Make a list of the reasons under the headings 'Push' and 'Pull'. The 'push' factors are those which made them want to move away from wherever they were, and the 'pull' factors are those which made them want to move into a particular area.

In surveys such as these, which involve people's private lives, you do not keep a record of any names, only of total numbers. When you have found out the information, make a map to show the moves. You will need to know where they moved from and to for these maps, but only the general areas, not the actual addresses. The moves can be shown as straight lines, with an arrow to show the direction of the move. You may find that you need more than one map. For example, to show local moves you may need a large-scale map of your local area. To show moves between regions you will need a smaller-scale map showing the whole of the UK. To show migration into the UK you may need a world map to show the countries of origin.

2 Arrange a class 'role-play simulation' of the Drumbuie inquiry.

In this type of simulation, each person, or a small group, plays the part of one of the members of the inquiry. The suggested parts, or 'roles', are:
(a) a local resident in favour of the scheme; (b) a local resident opposed to the scheme; (c) the representative of one of the large oil companies; (d) the representative of the platform-building companies; (e) the representative of the Highlands and Islands Development Board; (f) the representative of the Scottish Tourist Board; (g) the representative of the Wester Ross County Council; (h) the Official Recorder, who hears all the evidence and who sends a report to the Minister to enable him to make a decision. The idea is that you try to convince the Official Recorder that your case is the one that should be accepted. It may be a good idea to have the 'Official Recorder' played by a panel of judges, rather than one person, who between them can decide who presented the best arguments.

You should do some 'homework' for your role. Find out as much as possible about the views of the people you are playing and as many facts as possible to support your case. The local library may be able to help you in finding information, but some possible sources are:

(a) The national press, particularly the *Guardian, The Times*, the *Daily Telegraph*, the *Sunday*

Times and the *Observer*, between about October 1973 and August 1974

(b) The National Trust for Scotland

(c) The Petroleum Information Bureau

(d) The big oil companies (Shell, BP, etc.)

(e) Mowlem and Taylor Woodrow, who had hoped to build the platforms at Drumbuie

(f) The Highlands and Islands Development Board

(g) The Wester Ross County Council.

3 (a) Arrange an exhibition of materials which show the value of immigrant communities to the UK. If you need help, arrange for one of your class to write to the Community Relations Commission, 15–16 Bedford Street, London WC2E 9HX, who may be able to help, either by providing materials or by suggesting where they might be found.

(b) It may be possible to arrange a class conference on the problems of minority groups with, perhaps, a speaker invited from the Community Relations Commission.

10 Your move

Figure 170 Diagram to show interlinked networks

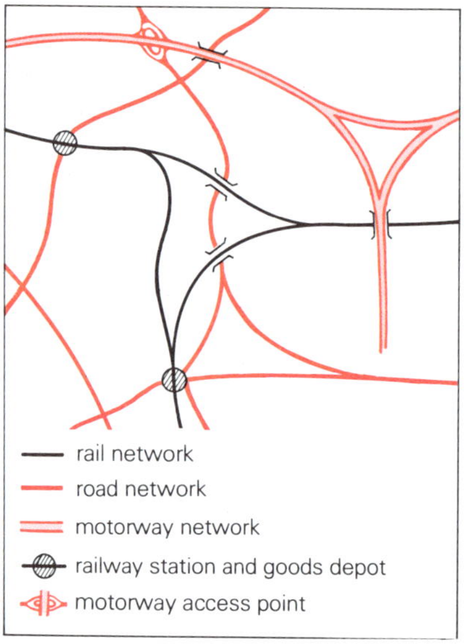

(a)

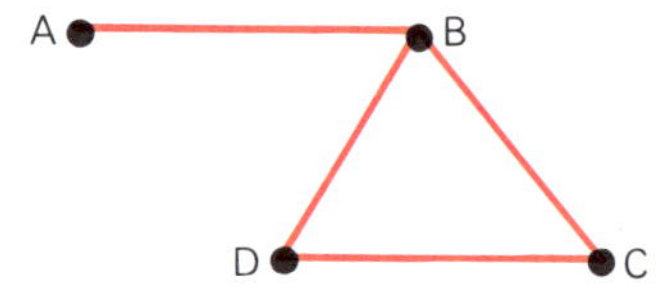

the numbers in the boxes show the number of links taken from each place to each other place

(b)

	A	B	C	D
A	—	1	2	2
B	1	—	1	1
C	2	1	—	1
D	2	1	1	—

Figure 171 (a) An imaginary transport network (b) Accessibility matrix

Posting a letter

You have probably posted a letter recently, since some 40 000 000 are posted every day. Have you ever thought about what happens to the letter after it leaves your hand? If it is being sent to someone a long way away, there are several different ways in which it could travel. One typical system would be:

(a) Collection by van from the post box. Taken to sorting office.

(b) From the sorting office by van to the railway station.

(c) From the station, by train, to another station near to the destination.

(d) From this station, by van, to another sorting office.

(e) Delivery from this office to the letter box by a postman on foot or on a bike.

There are, of course, many variations on this pattern. For instance, if you were writing to one of the Hebridean islands mentioned in the previous chapter, the letter might also travel by boat and air.

All the ways in which the letter travels are **means of transport**. The roads, railways, air and sea lanes are **transport routes**. It may be that you think of them as 'lines'. Although this is true to some extent, there is more to it than that. To travel from one part of the country to another by, say, rail, you often have to change trains. This is because one line has to link up with another. A system of lines linked together like this is called a **transport network**. Each means of transport has its own network and each network links up with the others at stations, ports and airports. In this way, there is a total national network.

Accessibility again

Much can be learnt about the usefulness of transport networks before even looking at the system in the

Figure 172 (above, left) *The same network as in figure 171, with the links labelled*
Figure 173 (above, right) *Links matrix for figure 172*

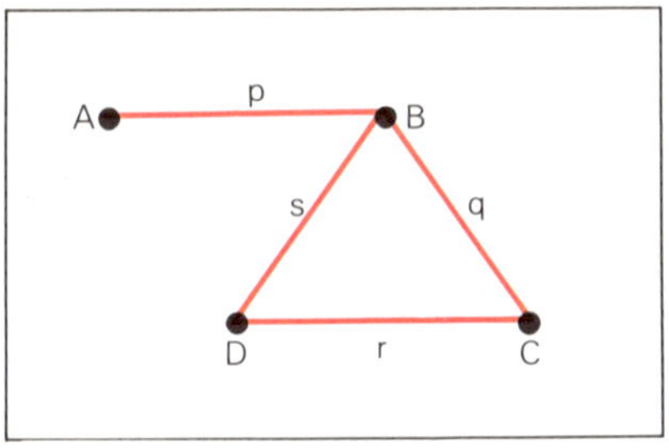

link	number of uses	total
p	~~////~~ /	6
q	////	4
r	//	2
s	////	4

UK. The section on accessibility on page 72 explained how to work out the importance of places by counting the number of routes entering them. In figure 171 the transport routes are shown by straight lines and the towns by dots.

1 Place the towns in order of accessibility, in the same way as was explained for figure 99 on page 72.

There are other ways of measuring accessibility in a network. If you think of the lines between places as links, you can count the number of links it takes to get from one place to another by the shortest route. In figure 171 to get from A to B takes one link; to get from A to C takes two links. Figure 171(b) shows all the links to get from each place to each other place on an 'accessibility matrix'.

Improving a network

2 You can try looking at networks in your school if it is one with several corridors on one floor. Make a plan of the corridors, with each of the junctions shown as a dot and the corridors between the junctions as links.

Work out which is the busiest link. Check your answer by seeing how many people use each corridor during lesson changeovers.

If your chosen link does not turn out to be the one which is actually the busiest, can you think of any reasons for the difference?

One way to find which is the busiest is to see which one is used most in getting from each place to every other place. To do this, you need to label the links. You can now draw a links matrix. For each

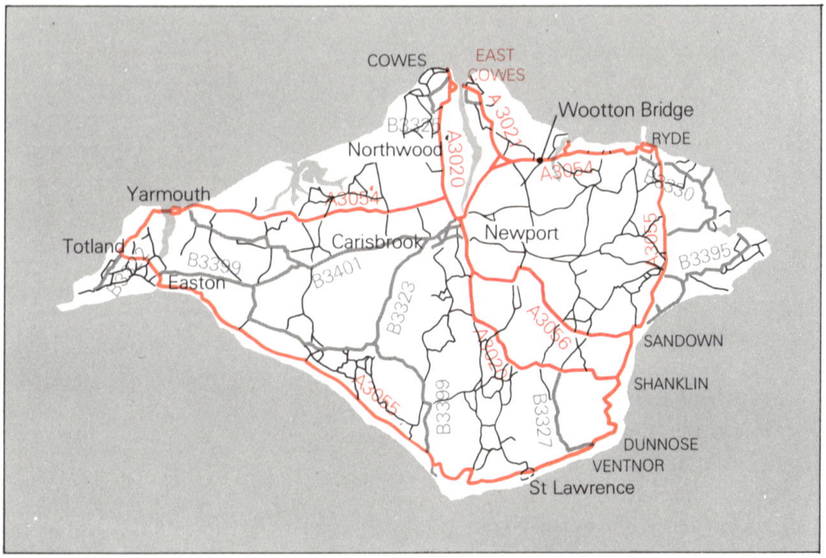

Figure 174 (below) *The Isle of Wight*

possible journey, from one place to all the others, tick off on the matrix which links have been followed. For instance, on figure 172 to get from A to C you use links *p* and *q* so they each get a tick. The completed matrix for figure 172 is shown in figure 173. Link *p* is the busiest.

The Isle of Wight network

This way of analysing networks can be taken further. The Isle of Wight (figure 174) is an important holiday area and the free flow of traffic along the roads is essential. Suppose you wanted to improve the roads on the island, but you had only enough money to do one link. Which link would you choose? One way of deciding would be to choose the busiest.

3 Find out which would be the busiest link on the Isle of Wight.

There may be more than one answer to this, because in some cases there is more than one shortest path between places and it depends which one of these you choose.

It will probably make it easier for you if you redraw the map. Make the roads into straight lines and the towns into dots, as on figure 171. There are three road junctions on this map which are not towns. These will have to be shown as dots on your redrawn map, for the purpose of counting links. Thus, from Newport to Ryde there will be two links.

You should not put these junctions on your matrix, they are simply for counting links. The matrix just needs the names of the towns along the top and up the side (figure 175(b)).

Flows along the network

Counting the links works only if the amount of traffic between each of the places is the same, which is unusual.

Figure 175 (a) Part of the Isle of Wight main road network redrawn as needed for question 3 (b) Partly-completed accessibility matrix

(a)

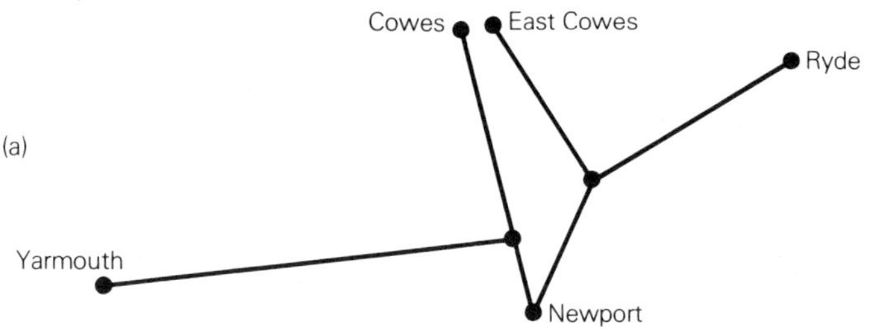

(b)

	Cowes	Newport	E. Cowes	Yarmouth	Ryde
Cowes	—	2	4	2	4
Newport	2	—	2	2	2
E. Cowes			—		
Yarmouth				—	
Ryde					—

You can experiment with size in the school corridor network. The rooms which hold the most people in the normal life of the school will also tend to make the most traffic, both going into and coming out of the room in question. There may not be a direct exchange between the two largest rooms, yet each will contribute to the flow.

The flow does not have to be of goods or people (figure 176). It may be in the form of telephone messages, or even radio broadcasts, but in each case the flow is from a place which has something to a place which wants it. On pages 36 and 37 in chapter 3, we saw that the power stations of the UK are to be found where there are suitable location factors. From the power stations, there is an outward flow of electricity to the factories and homes which use it. Naturally, the biggest flow is from the largest power stations to the biggest conurbations which use the most electricity. This general picture is often changed. Sometimes one part of the country needs more electricity than others; for instance, there may be a cold spell in one part of the country whilst another is having warm weather. This is quite a common thing in our varied climate.

Of course, flows do not go all one way. Usually, although one product, or message, goes in one direction, there is another product, or message, or payment of money, in the other. Since in any one direction the biggest flow is from the largest producer to the largest consumer, it follows that the maximum two-way flow is likely to be between the two largest places in an area (figure 176).

4 Between which two areas of the UK would you expect to find the biggest flow along transport lines?

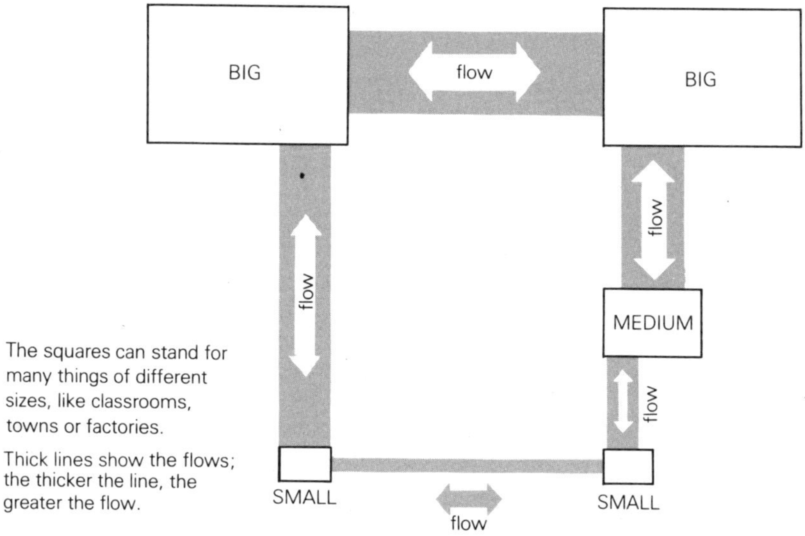

Figure 176 The creation of flows

Shortening the distance

Movement around the school may also illustrate the importance of distance as well as size. In general, movements from one classroom to another will be over as short a distance as can be arranged to fit the timetable. There will obviously be a saving in lesson time if this can be done.

The same is true for transport flows in the UK as a whole. There will be a saving in time and money if distances can be reduced. Usually, the further apart places are, the less traffic there is between them.

Forecasting the flow

One way of trying to estimate the possible traffic flows between places has been borrowed from the laws of gravity. For this reason, it is known as a **gravity model**. (See page 67 to remind yourself of what a 'model' is.) To find out how much traffic there will be between two places, you multiply the population of the two places and then divide by the distance between them. The result of the sum is just a number. It cannot mean numbers of cars, or tonnes of cement, so the word **interaction** is used to mean any kind of flow. Even so, it does show which parts of a network are the busiest.

A shorthand way of writing down the sum is as follows:

$$I = \frac{P_i P_j}{D_{ij}}$$

In this equation, I means the 'interaction'. P_i means 'the population of imaginary town i'. P_j means 'the population of imaginary town j'. If you write the two next to each other it means 'multiply'. D_{ij} means 'the distance from town i to town j. Thus the whole equation means, *The interaction is equal to the population of imaginary town* i, *multiplied by the population of town* j *and divided by the distance between towns* i *and* j.

5 Calculate the interaction between London and the West Midlands conurbation.
West Midlands conurbation population: 2 million
Greater London population: 7 million
Distance between the two conurbation centres: 187 km

6 Work out the interaction between the West Midlands conurbation and Merseyside.
Merseyside conurbation population: 2 million
Distance from Birmingham to Liverpool: 145 km

7 How do you account for the difference in the interaction in questions 6 and 7?

8 Work out the interaction between any other two places listed in figure 91. Look up the distances in an AA book.

9 (a) Work out the interaction between all the conurbations listed in figure 91.
(b) Make the result into a flow diagram. Draw a map of the UK showing each conurbation centre as a dot or small circle.

Draw straight lines to join all the conurbations to each other, but make the thickness of each line proportional to the interaction. A suitable scale would be one millimetre of line width to 10000 units of interaction.

This gravity model is a very crude one. But estimating the traffic of the UK is very important, and so models are used. In looking at real-life problems, the models are very much more complicated, needing computers to work them out, but the ideas are similar to the ones you have used.

Transport costs

Forecasting is vital in planning new routes and it is also important in the normal business life of the nation. Since increasing distance reduces the likelihood of demand for transport, how do the transport firms try to counteract this? If you own a lorry, it is earning money for you only when it is actually carrying goods. Therefore you would like the lorry journeys to be as long as possible. But the need for transport decreases with distance.

10 Before reading on, see if you can think of any ways in which a transport company could persuade customers to use longer journeys more often.

What usually happens is that the freight operators charge slightly less as the distance of a journey increases. For instance, to carry one tonne of goods 100 km the cost might be, say, £10. But to carry the same goods 200 km the charge would not be twice as much. Perhaps it would be, say, £18. Then for 300 km it might be £23. This method of charging is called 'tapered freight rates'.

For anybody using the system, the effect of distance is reduced as far as costs are concerned.

'Time distance'

Although cost is always important, in many cases time is vital. Perishable goods, such as fresh fish or strawberries, would lose their value if they spent a long time on the journey. Passengers, too, may prefer a more expensive fast journey to a cheaper slow one. Therefore, the effect of distance in reducing traffic can be counteracted by using faster means of transport. You can even think of places getting 'closer' as a result of changes in transport.

11 Using figure 177, make a map of the changes in travel times in Britain. Trace the towns and the lines connecting them to London in figure 178 but do not include the coastline. The length of the lines at present shows the distance of the towns from London, if you use the scale.

Now, make the length of the lines proportional to the *time* it takes to get to each town from London. You will have to choose a suitable scale.

When you have drawn the lines to the right length for each town, draw in the coastline as accurately as you can, freehand.

Next, do the same exercise, but from the train timetable. Use the same scale.

You could try to do the same exercise for the air timetable, even though not all the places are on it. Unless you use a very big piece of paper for the first exercise, you will find that on the same scale, this third map is so small that you cannot draw it.

You may find that the same ideas affect your own behaviour. If you travel to shop, or for entertainment, perhaps you usually go to a place which is a greater distance away than another which you could use, but 'closer' in time because there is a direct bus or train route.

London to	Distance (miles)	Journey Times (hours) Stage coach	Motor coach	Rail	Air
Aberdeen	524	–	–	10	–
Aberystwyth	234	–	–	5½	–
Bradford	189	–	5	2	1
Brighton	50	7¾	2	1	–
Bristol	118	12¼	4¼	1¾	–
Cardiff	145	17½	6	2¼	–
Carlisle	300	32¾	8½	4½	1½
Edinburgh	393	43	11	6	1¼
Exeter	170	18¾	7¼	3	–
Glasgow	401	42½	12	6¾	1½
Hull	197	18½	3½	4	1
Leeds	180	21½	4½	2½	1
Manchester	183	19½	5½	3	¾
Newcastle	268	30¼	8½	4	1
Nottingham	125	12¾	4½	1¼	–
Norwich	115	12	5	2¼	–
Sheffield	156	17	4	1½	–
Southampton	79	9	3¼	1¼	–

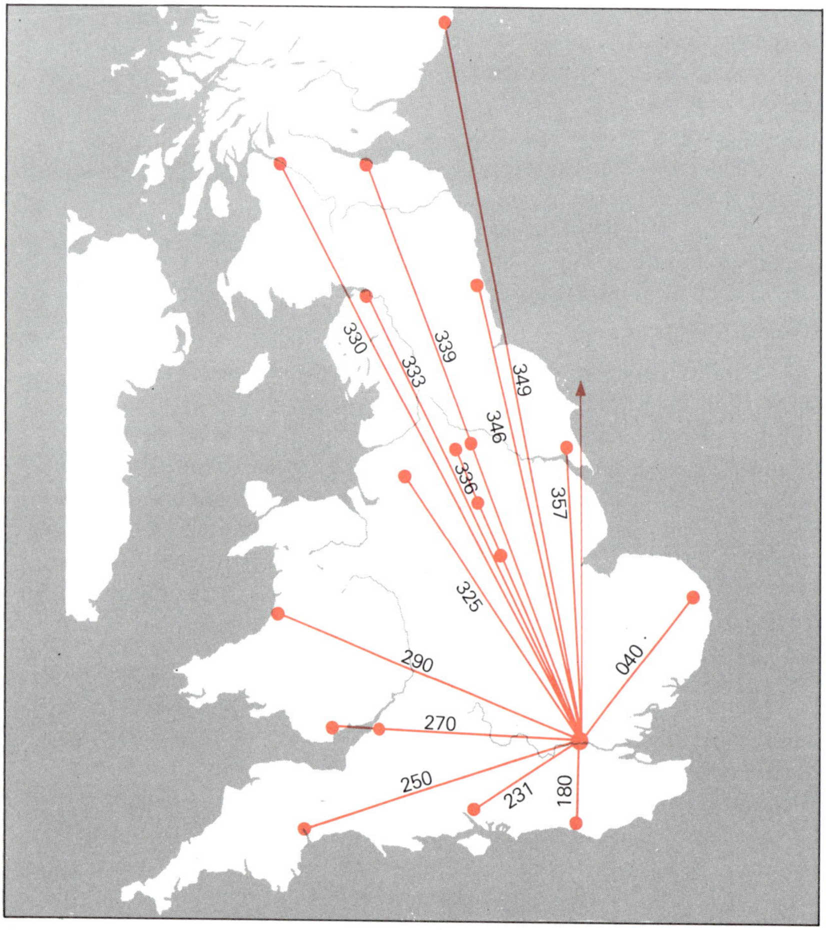

Figure 177 (left) Journey times
Figure 178 (below) Great Britain, showing the bearings to the towns named in figure 177

Motorways

In a similar way, the growth of the motorway network has revolutionised motor transport. It is not so easy to compare times between different places because most forms of motor transport do not operate to a timetable. However, it used to take a family motorist, travelling at a fairly normal speed, about five hours to get from London to Sheffield. Now, it takes a little over three hours. The motorway network of Great Britain has been built up piece by piece to link the conurbations. Now that the main inter-city motorways have been completed, more crowded, or congested, areas are being linked to them by shorter lengths to relieve the pressure of the traffic (figure 179).

It might be thought that the motorways were entirely for the good of the nation. They do, after all, allow for fast movement of both goods and people, without the need to travel first to a railway station. The movement of goods is vital to the economy and the motorways are a most important contribution to this movement. Even so, all is not entirely well. As with almost any modern activity there is conflict. There is an obvious conflict over land use. Imagine how much farming land is taken up by a motorway intersection or service area.

In urban areas, the sheer size of motorways and their interchanges takes up space which is more valuable than agricultural land. The cost of the land, plus the cost of demolishing factories or houses to make room for the road, is so high that motorways are only rarely constructed into, or through, cities. The motorway 'box', which was planned

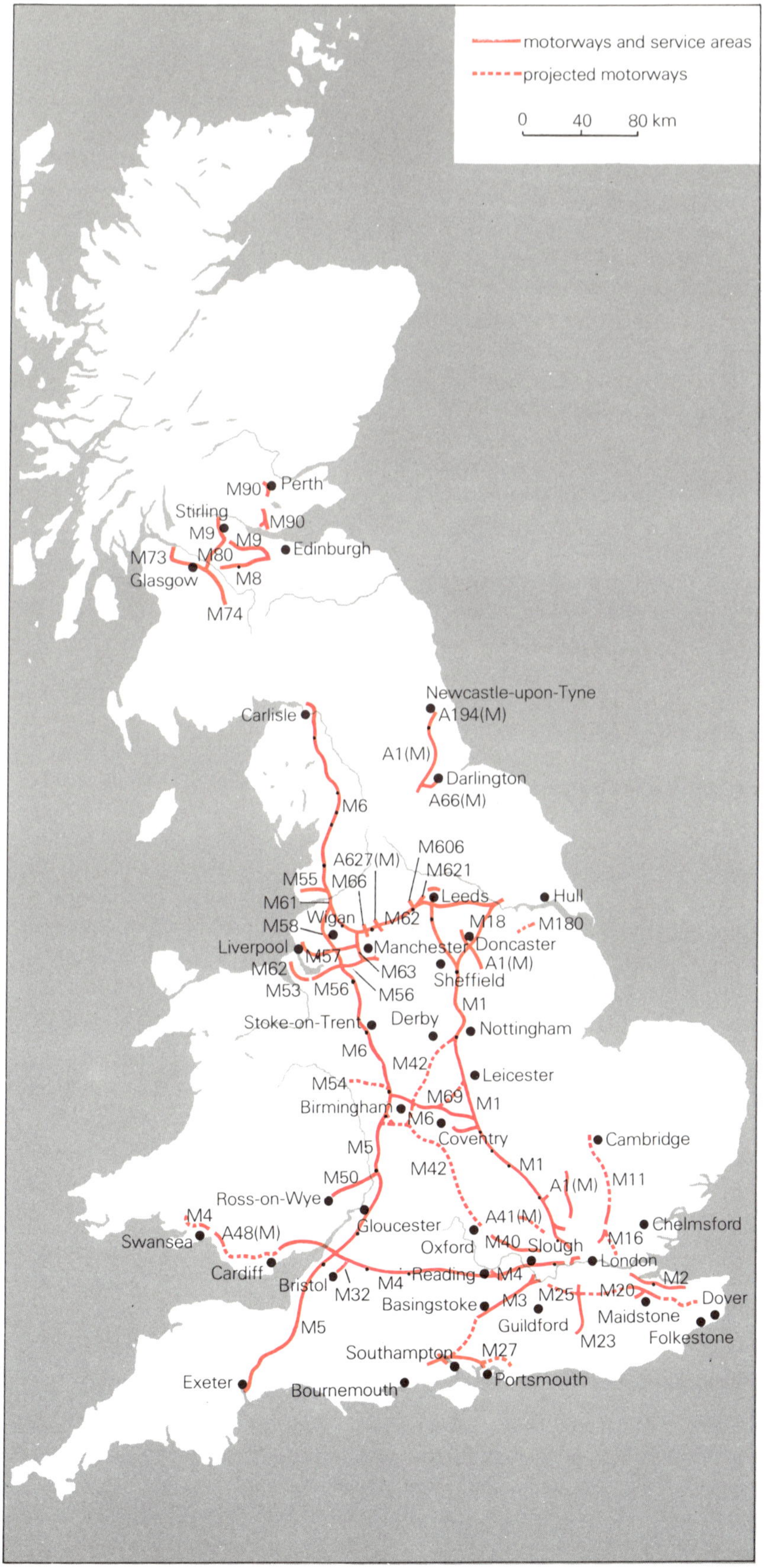

Figure 179 The motorways of Britain

in 1969 to speed traffic around inner London, was eventually scrapped for these reasons, and because of the enormous number of houses that would be lost at a time of housing shortage.

There is an even more obvious conflict over the noise and pollution in the areas through which motorways pass. How would it feel to be living in one of the flats in Hendon, shown in figure 180?

Motorways can cause such a nuisance that there have been many instances of people making their feelings known, such as in figure 181.

Motorways are not very welcome in National Parks. Can you suggest why this might be? One of the less obvious reasons is that the parks are now much 'nearer' the centres of population than they were before the motorway age. The Lake District is now within easy day trip distance of the South Lancashire, Merseyside, West Yorkshire and even the West Midlands conurbations. Similarly, the Brecon Beacons Park is now 'not far' from the South East and from the West Midlands conurbation. Although the National Parks were *intended* for visitors, those near the motorways are now getting more than they bargained for.

12 Why are the words 'nearer' and 'not far' in inverted commas in the previous paragraph?

Even where the motorway itself does not cause controversy, there may be another problem. Because of their convenience, motorists naturally tend to use motorways rather than other roads. This means they have more vehicles on them. They are said to 'generate' their own traffic. But what happens when this extra traffic reaches the end of the motorway? Vehicles are arriving faster than they can get away. Con-

Figure 180 Fiveways interchange, Hendon

Figure 181 Demonstrations over the opening of Westway – the longest elevated motorway in Europe

gestion occurs. In trying to escape, back streets are used. And so, what was a quiet residential street becomes clogged up with unwelcome visitors. Carefully planned traffic schemes may help to reduce this, but it is difficult. As one planner said, 'Even if you extend the motorway, you never cure the traffic jam, you only move it on a bit.'

13 Why do you think there should be a demonstration at point B in figure 182 about a motorway which has been opened at point A?

14 Find on the map the places named in the newspaper article.

15 In your own words, explain what you think are the fears of the Chiswick residents, and what they would like the government to do to make their stretch of road safer.

Planning motorways

In deciding on a motorway network, the planners have to try to build the most convenient system at the lowest possible cost. Once the main lines of the network were agreed, the detailed route of each motorway would have to be planned.

At this stage, some of the factors to be considered would be:

(a) Ease of construction, which would affect cost. Steep gradients, big cuttings or embankments, demolition work, building bridges over rivers or other roads, are all things which would be kept to a minimum.

(b) Cost of buying the land. Urban areas and rich farmland would tend to be avoided if possible.

(c) Disturbance. Thickly populated areas, even outside towns, would be avoided if possible.

(d) Effect on the landscape. Where it could be done, the motorway route would be kept clear of areas of great scenic beauty.

Naturally, not all these conditions can be met all the time. In some cases, the cost of the road is greatly increased. In others, there is strong opposition from people who would be affected by a new motorway. The final decision usually tries to help everyone a little, but does not go the whole way to solving all the problems – it is a compromise.

16 At some time in the not-too-distant future, the construction of a motorway is suggested to link the M6 near Birmingham with the M1 near Nottingham. The general route lies across OS 1:50000 sheets 128, 129, 139, 140 or one-inch sheets 120, 121, 131, 132.

If you have these sheets in your school, a group of you could try to plan the detailed route of the suggested motorway, bearing in mind the factors listed above.

Of course, you will not have all the information that the planners would have. You will have no precise knowledge of building costs, for instance. But from the map you

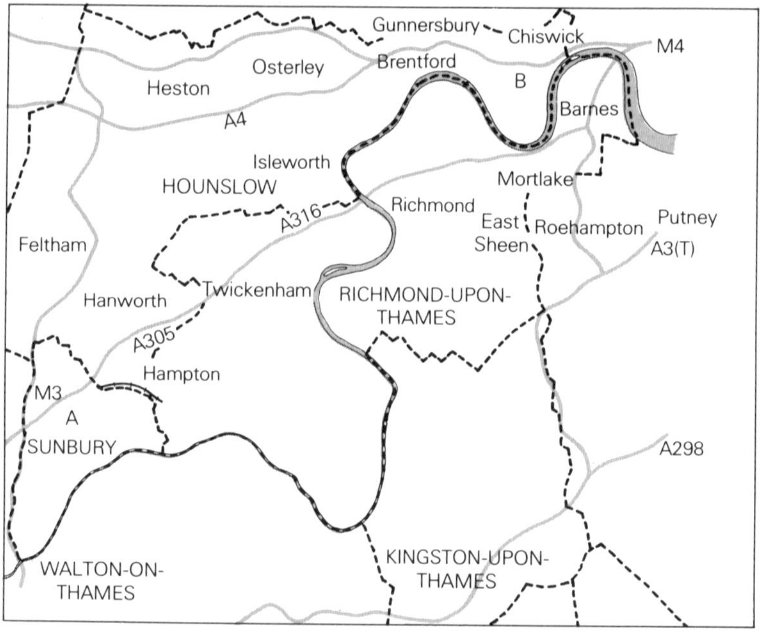

Figure 182 The opening of the M3

TRAFFIC STOPS FOR M3 DEMOS

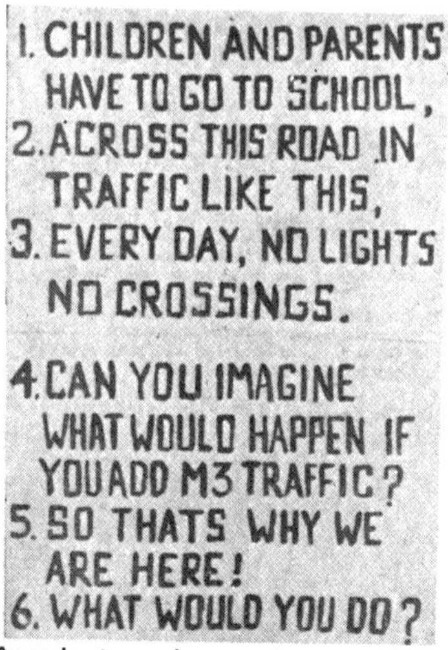

An explanatory notice to soothe the frustration of irate drivers, stuck in a jam that tailed back for almost a mile to the Richmond border

ACTION on the A316-M3 issue continue last week on several fronts with road ing demonstrations, questions in Parli and a council delegation all seeking t surise Fred Mulley, Ministe into delaying the motorway'

Among the successes Department of the Envi junctions with Edensor R are likely to be installed cording to a G At the two-ho tween Fred Mull and a Hounslow tion, i Barn it w th

could make a good estimation of the other factors: gradients, number of bridges, built-up areas and so on. Estimating the scenic beauty from the map is more difficult, but sometimes possible. Even if you can decide on what the scenery is like, your group may not be able to agree on a policy towards it.

Connections

In any network there are a number of links and a number of junctions or terminals. The smallest number of links that a network can have is one less than the number of junctions. Terminals have to be counted as junctions.

In figure 183 if you divide the number of links by the number of junctions, you get an answer of 0·75. If you add one more link, the answer is 1, because the number of links and junctions is the same. From the point of view of the user, the network is better connected because you would be able to get directly between two more places. If you add one more link, the answer to the division sum becomes 1·25 and the network is still better connected. However, you still would not be able to get directly from every place to all other places.

This method of showing the **connectivity** of a network is called the **Beta Index**. It can be written down in a shorthand way as:

$$\frac{l}{j}$$

Figure 183 Four junctions, three links

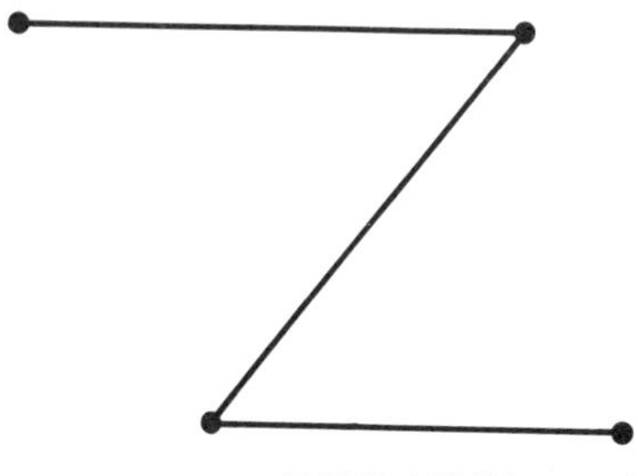

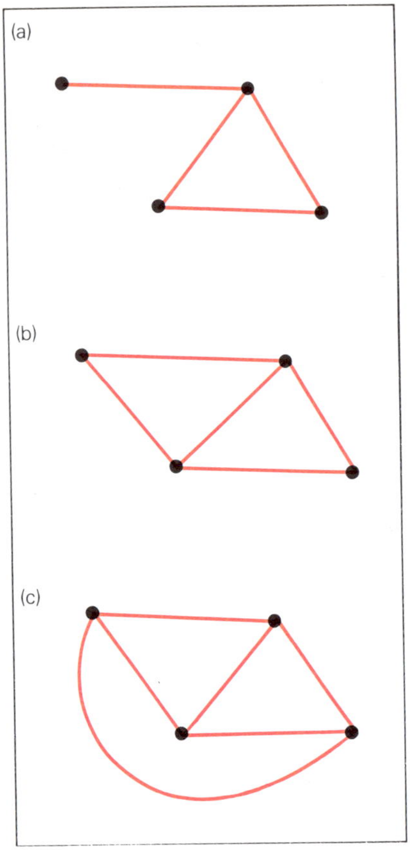

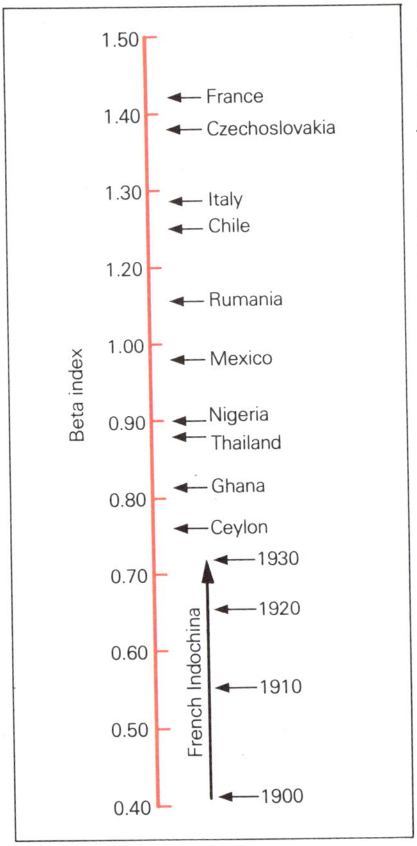

Figure 184 (far left) (a) Beta Index 1·0 (b) Beta Index 1·25
Figure 185 (left) The Beta Index for the rail networks of selected countries

with 'l' meaning 'links' and 'j' meaning 'junctions'. The j being placed under the l means 'divide l (the number of links) by j (the number of junctions)'.

17 What is the Beta Index of the third network in figure 184?

The higher the Beta Index, the better connected is the network. Figure 185 shows the Beta Index for the rail network of certain countries.

18 Is there any difference between the Beta Index values of more developed countries and those of developing countries?

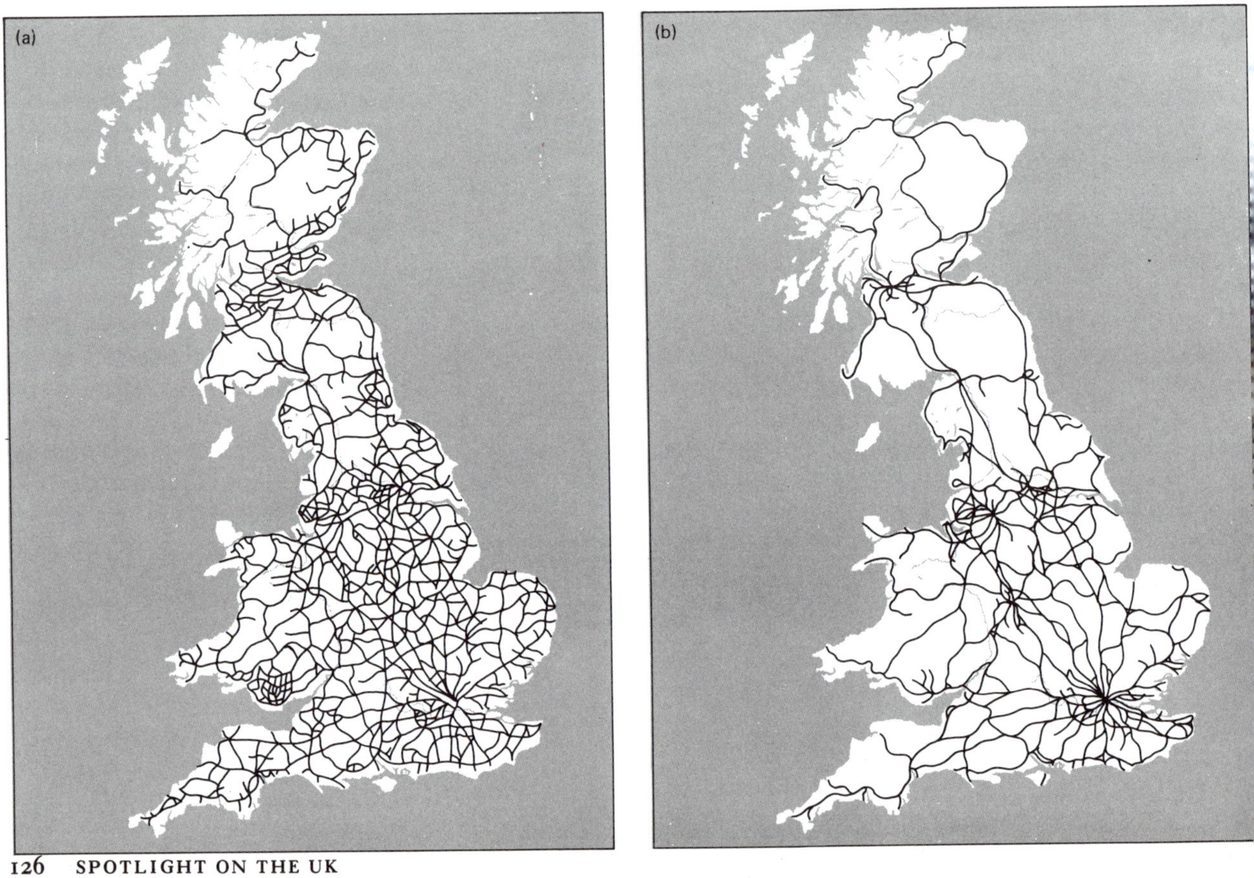

Figure 186 (a) The rail network of Britain in 1920 (b) The rail network of Britain in 1974

19 Roughly where would you expect the UK to come in this 'league table'?

Although the Beta Index is useful, it does not show what the maximum number of links is for any particular number of junctions. If you go on adding links to a network, you will eventually have more than one link between the same two places.

Pruning the railways

An example of using this kind of analysis is to be found in the thinning out of Britain's rail network. In the 1950s, the railways were becoming uneconomic. In 1961 the government brought in Dr Beeching to try to find a way round the problem. His proposals, in the 'Beeching Report' of 1963 were to close over half the length of track in Britain. One of the reasons for this was that there were more links than junctions. Therefore, there were often two lines where one would have done. That is to say, many stretches of track were 'duplicated'. The duplications were mainly between places which offered the greatest profit to the railway companies when they were built in the nineteenth century.

20 From your knowledge of the gravity model (pages 120 and 121), which would have been the most profitable routes in the nineteenth century?

21 Compare figures 186 (a) and (b). In which areas of Britain, other than between the great industrial centres, have railway lines been closed?

Some lines were built in rural areas because at that time they were the chief means of passenger transport. Since the Second World War, the number of motor vehicles has increased enormously, from 3·5 million in 1947 to 14·8 million in 1969 to 16·1 million in 1972. As a result, the railways now carry only a small part of the passenger traffic of the UK.

Even over the last ten years, the process has continued. Figure 187 shows the numbers of miles which passengers were carried in different years. You can see that it is several thousand million in almost all cases.

22 How many more miles were passengers carried in 1972 than in 1962
(a) by air?
(b) by private transport on the roads?
What happened to rail passenger transport during the same period?

23 (a) What percentage of the total was carried by rail in 1962?
(b) In 1972?

For reasons such as these, it was necessary to close down many of the 'mainly passenger' lines. The cost of keeping such railways running was far greater than the income from fares. Although such closures were needed to make the railways more economic, they were not so good socially. Although there are so many cars, not everyone has a car. There were many objections to the 'Beeching axe' because it caused real hardship to a large number of people. In fact in many areas, uneconomic lines are still kept going by government grants. The *most* uneconomic lines *have* been cut. At present only about half the track which existed in 1948 is used for passengers. The grants are used only for those lines which the government thinks most need them.

One area in particular needs such help. This is the London commuter belt. Here, a vast number of people move in to work in the morning and out again in the evening, in a 'tidal flow'. During the rest of the day and night, the lines are little used. To buy and maintain the equipment and buildings and to employ the staff needed during the rush hour, costs far more than it is possible to pay for by fares. But this commuter network is so important, both to the people who use it and to the working life of London, that it is kept going by grants.

Figure 188 is a flow diagram to show the numbers of people travelling by British Rail into London during the morning peak period.

24 Judging by the flow diagram, from which parts of the South East do most commuters come?

25 At which station do most arrive?

Despite the large numbers carried in the South East, in general there is a changeover of passengers from rail to road, as we saw in figure 187. The same thing is happening with goods

Figure 187 Passenger transport in Great Britain: estimated passenger mileage. Source: Department of Environment

Thousand million miles

	1962	1963	1964	1965	1966	1967	1968	1969	1970	1971	1972
Total	173·0	180·2	196·3	206·7	216·7	228·0	236·6	243·1	254·4	269·4	278·8
Air	0·7	0·8	0·9	1·0	1·1	1·2	1·2	1·2	1·2	1·2	1·3
Rail	22·8	22·4	23·0	21·8	21·5	21·2	20·8	21·6	22·2	22·0	21·1
Road:											
Public service vehicles	42·4	41·5	40·3	39·2	38·2	37·7	36·9	36·3	34·8	34·8	34·2
Private transport	107·1	115·5	132·1	144·7	155·9	167·9	177·7	184·0	196·2	211·3	222·2

Figure 188 The flow into the centre: numbers arriving at British Rail stations during the morning peak

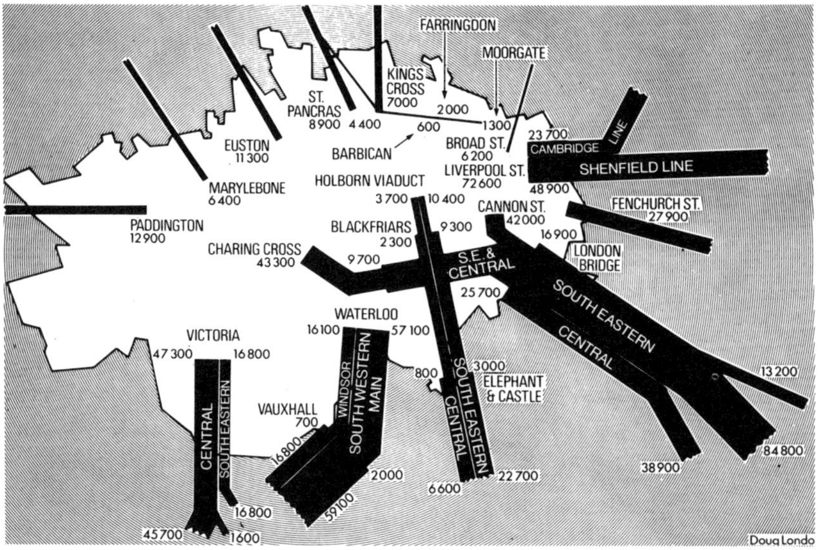

transport, only not to quite the same extent.

In figure 189, tonne-miles are similar to passenger-miles. In this case it means the total number of miles that every tonne of goods has been carried. It is a useful way of measuring different types of transport.

26 (a) What percentage of goods traffic was carried by rail in 1962?
(b) To what percentage had this changed in 1972?

27 (a) What percentage was carried by road in 1962?
(b) In 1972?

Can the railways survive?

The reasons for the change are not hard to find.

28 In figure 187, which means of transport is used by the largest number of passengers?

By using their own cars, people can travel from door to door. Even using public transport, most people do not have to travel as far to a bus stop as they do to a railway station. In a similar way, a lorry can travel on minor roads, on trunk roads and on motorways, all without stopping to unload and reload. Such door-to-door collection and delivery is not possible by rail unless a factory has its own sidings. Road transport is said to be more **flexible**. From all this, it would seem that the railways are doomed, yet this is by no means the case.

Figure 190

To start with, the roads have many disadvantages. Everyone knows of the nuisance caused by 'juggernauts', particularly when they use the minor routes which make the road network so flexible (figure 190).

Figure 189 Goods transport in Great Britain
Source: Department of Environment

	1962	1963	1964	1965	1966	1967	1968	1969	1970	1971	1972
Total tonne-miles (thousand millions)	64·6	66·0	72·2	73·7	76·1	75·5	79·5	81·1	82·5	80·5	80·6
Road	33·6	35·1	40·2	42·1	44·8	45·6	48·3	49·3	50·0	50·4	51·4
Rail	16·1	15·4	16·1	15·4	14·8	13·6	14·7	15·3	16·4	14·9	14·2
Coastal shipping	14·3	14·9	15·1	15·3	15·5	15·2	15·0	14·8	14·2	13·1	13·0
Inland waterways	0·2	0·1	0·1	0·1	0·1	0·1	0·1	0·1	0·1	0·1	0·1
Pipelines	0·4	0·5	0·7	0·8	0·9	1·0	1·4	1·6	1·8	2·0	1·9

Site: Bridge, Kent.
Pop: 762

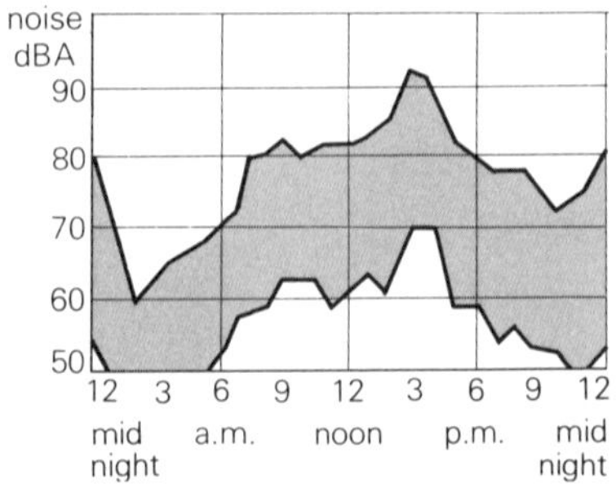

A small village on the main road (A2) to Dover, battered by the continual passing of traffic: lorries sometimes have to mount the kerb in the centre of the village to squeeze past each other. Many of the houses are less than six feet from the road.

What the people said
Of the 39 people interviewed, 16 had some difficulty with holding a conversation, or listening to the TV or radio, and 15 had some difficulty sleeping because of the noise.

Our findings
Our measurements show that the daytime noise-level was very high, and the residents only gained a brief respite between 1 and 5 a.m.

Site: Westway, London.

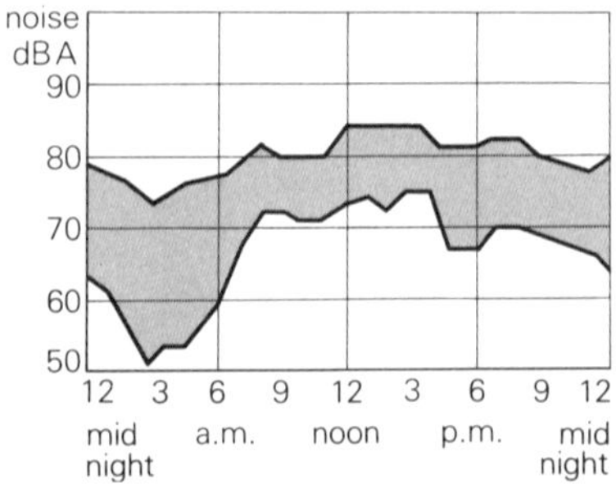

One of the main arteries in and out of West London – a dual carriageway section of the A40. Forty m.p.h. speed limit. Terraced houses each side, about 15 feet back from the kerb.

What the people said
Of the 48 people interviewed, only a third found that the noise of traffic could interfere with conversations or listening to TV or radio, though over half had some difficulty sleeping.

Our findings
Our measurements show that this road was the noisiest of our sites. Ironically, the huge volume of traffic, and the resulting constant daytime noise, seemed to annoy the people rather less than the varying levels of noise in other sites.

Site: Appleby, Westmorland.
Pop: 1946

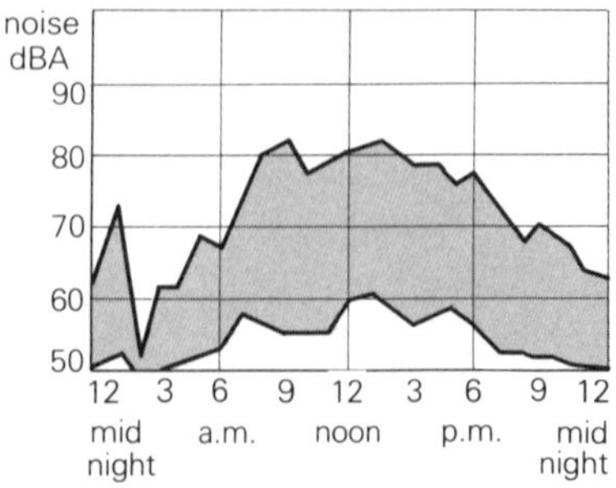

A fair-sized town on the A66, often used as a link route between the M6 and the A1. Most of the houses on the main road are within ten feet of the kerb.

What the people said
Of the 36 residents questioned, half found conversation or listening to TV or radio difficult. Over a third had some difficulty sleeping.

Our findings
Our measurements show a fairly constant high level of noise between about 8 a.m. and 6 p.m.

If the shaded section of the graph is broad, it means that the traffic noise is fluctuating a lot, depending on what traffic is passing at any given moment. A narrow shaded section means that the noise level is fairly steady all the time – in the daytime, this usually means that a lot of traffic is passing.

Our tests tended to confirm that a fluctuating noise level (broad shaded sections on graph) was more disturbing to people than rather higher but steadier noise level (narrow shaded sections).

Figure 191 An extract from Motoring Which, *April 1973*

In figure 191, the graphs may not be easy to interpret at first, but their purpose is made clear by the text in the 'box' at the foot of the extract.

29 Find in an atlas the places named in figure 191. Make a list of the difficulties that the people experienced because of traffic.

30 Which place do you think would be worst to live in because of traffic conditions?

Look particularly at the information for Bridge, Kent. That is where the photograph in figure 190 was taken.

The congestion of road traffic in towns and industrial areas is also well known.

Another reason why the railways can survive is that they still have some advantages over roads, in addition to creating less environmental damage. We have seen

Figure 192 (above, left) *A 160 k.p.h. Inter-City express on the London to Glasgow route*
Figure 193 (above) *A Freightliner train*

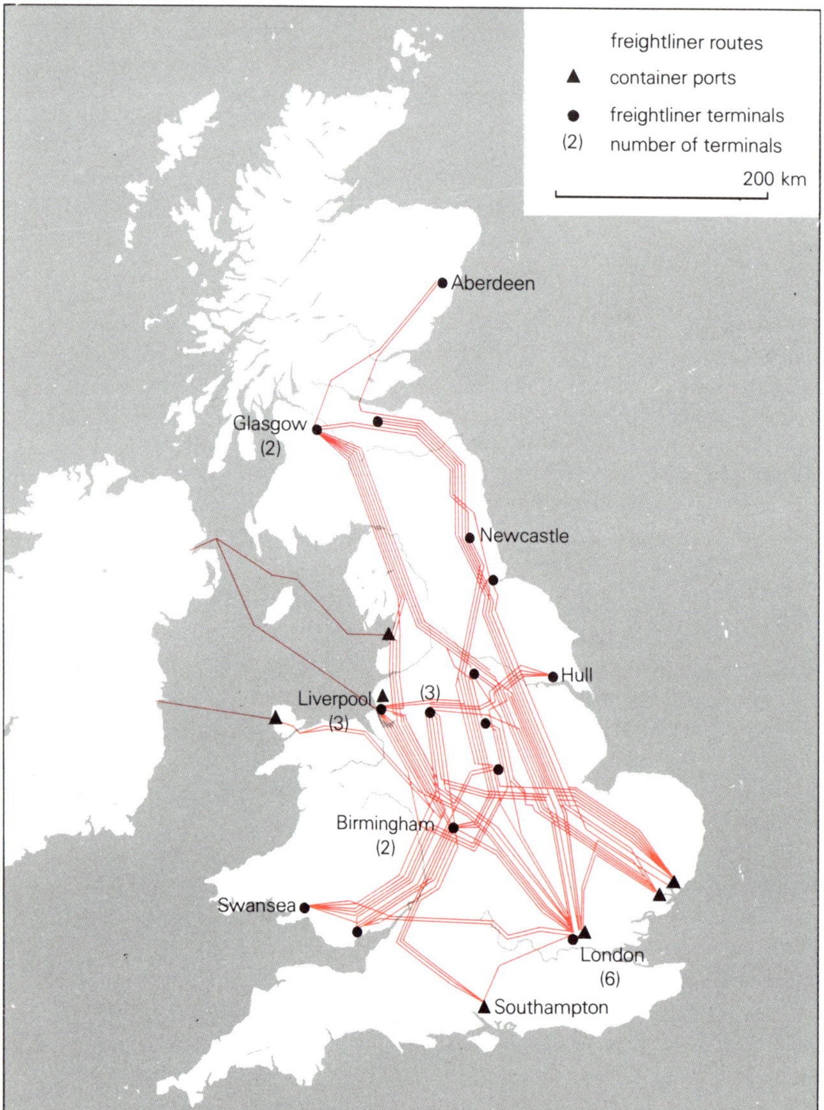

that speed is important, particularly for passengers. Over long distances, trains are faster than motor vehicles. Therefore, in order to compete, British Rail have introduced new, faster, services between the big cities.

The lines from London to the Midlands, the North West and to Scotland have been electrified and allow speeds of 160 kph (100 mph). To Wales and the West there were high-speed diesel trains which were not quite so fast. These have been replaced by the new High Speed Passenger Train which can travel at 200 kph (125 mph) on the existing track.

The railways can carry bulky freight more easily than lorries and have increased this advantage by adapting to the needs of the customers. For big users there are now 'block trains' where the whole train is given over to one company. Often the wagons are specially designed for the customers' needs.

Figure 194 *Freightliner routes*

Even general cargo, which consists of many different items, is being moved efficiently by rail on the Freightliner system (figure 193).

The great advantage of these trains is that they use containers of a standard size and shape. You may have heard of the 'container revolution'. The idea is very simple and is not used only by the railways. Instead of loading cargo item by item into a train, on to a lorry or into a ship's hold, the goods are loaded into a container, which is a large box. From then on, the goods are not unloaded until they reach their destination. If the journey involves changing from train to lorry or ship, the container itself is unloaded and reloaded into the new vehicle. Since all containers are made to a standard size, specialised equipment can be made to lift them and carry them around. The vehicles in which they travel can all be made to the same standards, so that there is no difficulty in transferring the container from one type of transport to another, or from one country to another. The saving in time and labour costs is considerable.

The map in figure 194 may seem rather difficult to follow at first, but it is just another kind of map transformation, like the one you drew of the roads on the Isle of Wight.

This chapter contains many facts about transport in the UK. Some of the facts are contained in the text and others can be discovered by careful study of the pictures. We have seen how changes in technology have revolutionised transport in this country. There is no reason to suppose that these changes are going to slow down. The facts, therefore, will be out of date in a few years. Some may even be by the time you read this (can you find any that are?). But if you understand the processes, you will easily be able to understand the changing facts.

PROJECTS

1 One way of understanding the processes referred to in the conclusion is network analysis. Another way is to examine the *causes* of different types of traffic.

(a) In this chapter we have considered various types of 'flow'. Look back through the chapter and make a list of all the different sorts of flow. For example, electricity movement in the national grid and the delivery of goods to a factory.

(b) For each of the types of flow listed in part (a), make a second list naming the types of places where the flows start and finish.

(c) Now try to work out the kinds of problems which are most likely to affect each type of flow, for example, traffic congestion in town centres or the placing of unsightly pylons.

(d) For each problem, suggest possible solutions and explain the difficulties of putting them into action.

2 One kind of flow that takes place every weekend in the winter, as well as on a good many other occasions, is the movement of football teams and supporters to away matches.

Work out which teams have the least distance to travel (to league matches only) during the course of one season. For simplicity, use straight line distances.

Figure 195 An example of a partly-constructed matrix for the English Football League, 1976. On this matrix the distances are taken from an AA book. For your own matrix you could do the same, or measure the straight-line distance from your map.

	Arsenal	Aston Villa	Birmingham City
Arsenal	—	110	110
Aston Villa	110	—	0
Birmingham City	110	0	—
Bristol City	116	87	87
Coventry City	91	19	19
Derby County	121	43	43
Everton	197	90	90
Ipswich Town	75	155	155
Leeds United	190	109	109
Leicester City	97	41	41
Liverpool	197	90	
Manchester City	184	79	and so on.
Manchester United	184	79	
Middlesbrough	246	173	
Newcastle United	274	200	
Norwich City	111	156	
Queens Park Rangers	0	110	
Stoke City	147	43	
Sunderland	272	200	
Tottenham Hotspur	0	110	
West Bromwich	110	0	
West Ham United	0	110	
Total	2832	2004	

The way to tackle this project is as follows:

(a) Make a list of the teams in, say, the English first division.
(b) From an atlas, make a map showing the towns or conurbations in which the teams are based.
(c) Make a matrix, with the names of the teams along the top and down the left side.
(d) On your map, measure the distance from each town to every other town and enter it on your matrix. Where a town or conurbation has two or more teams, the distance to away matches in the same town will have to be entered as 0, for ease of measurement.
(e) Finally, add the numbers together down the columns and enter the total at the bottom.

3 Using the method outlined in project 2, compare the distances travelled by English first division clubs and Scottish premier league clubs. Remember that in the Scottish premier league, the clubs visit each other twice.

(a) Work out the distance travelled by the club that travels furthest a season in each of the two leagues.
(b) Work out the distance for the club that travels the least in each case.
(c) Work out the average distance travelled by all clubs for each of the two leagues. To do this, you add up the distance totals for each club into a grand total for the whole of the league, then divide by the numbers of clubs taking part.

4 Leaving aside football hooliganism, what do you think are likely to be the problems connected with the movement of football supporters to away matches every weekend?

5 Find out:

(a) whether it is possible
(b) how long it would take
(c) how much it would cost

to travel from your home town, or nearest town, to (i) London; (ii) Glasgow; (iii) Belfast; by each of the following means of transport:
(a) train (or train and boat); (b) bus and coach (or bus, coach and boat); (c) air.

6 Organise a class debate on the following motion: 'To prevent further environmental damage, most freight transport should be directed back to the railways'.

11 The big ship sails down the Alley Alley O

Seaports

We saw in the previous chapter that the Freightliner network links up with the ports that handle containers. Beyond the ports is yet another transport network, the shipping routes. Ports are perhaps the most important of the points where people and goods change from one type of transport network to another.

Like lorries or trains, ships are earning money only when they are actually carrying people or goods. When they are in port they are not earning; in fact, they are having to pay harbour dues and other fees. It is common sense that a short turnround time is a good thing for shipowners. The people travelling, or sending goods, will also be happier for less delay.

The 'container revolution' has already been mentioned. It has particular advantages for ports. General cargo used to take a great deal of loading and unloading. With a container, the whole thing is vastly speeded up.

1 In figure 196, can you see any equipment specially designed to handle containers?
2 Can you see the containers stacked on the deck of the ship?

The chief disadvantages of container ports are that they cost a great deal to build and equip, and the space for storing the containers has to be a lot bigger than the old transit shed.

In figure 197, the container storage space occupies 40 hectares, and the unused land behind it, which is available for future development, occupies 74 hectares.

Figure 196

3 To give you some impression of the size of the area, look again at figure 196 and compare the size of the cranes with the people. Then look at figure 197 and compare the size of the cranes with the extent of the storage area. Compare the area with the cars to the right of the container area in figure 197.

For handling containers many port authorities have built an entirely new port, in addition to the existing one, but not all ports can afford container ports immediately. However, they are to some extent a 'status symbol' for a modern port and more can be expected as money becomes available.

Such an efficient system seems to be a good one for all concerned. Yet, as with most new ideas, the changes involved create difficulties for some. The dock workers were highly skilled at loading and unloading ships, but large numbers of them were needed.

Figure 197 The container port, Southampton

The container system is much quicker than traditional turn-round methods and it needs many fewer workers, since it is highly mechanised. The dock workers were naturally worried about their jobs, but their fears have been partly overcome. Many men have changed their skills to operate the new machinery. Agreements have been made to ensure that needless unemployment or low wages do not happen. Further, many dockers are still needed in the older parts of the ports.

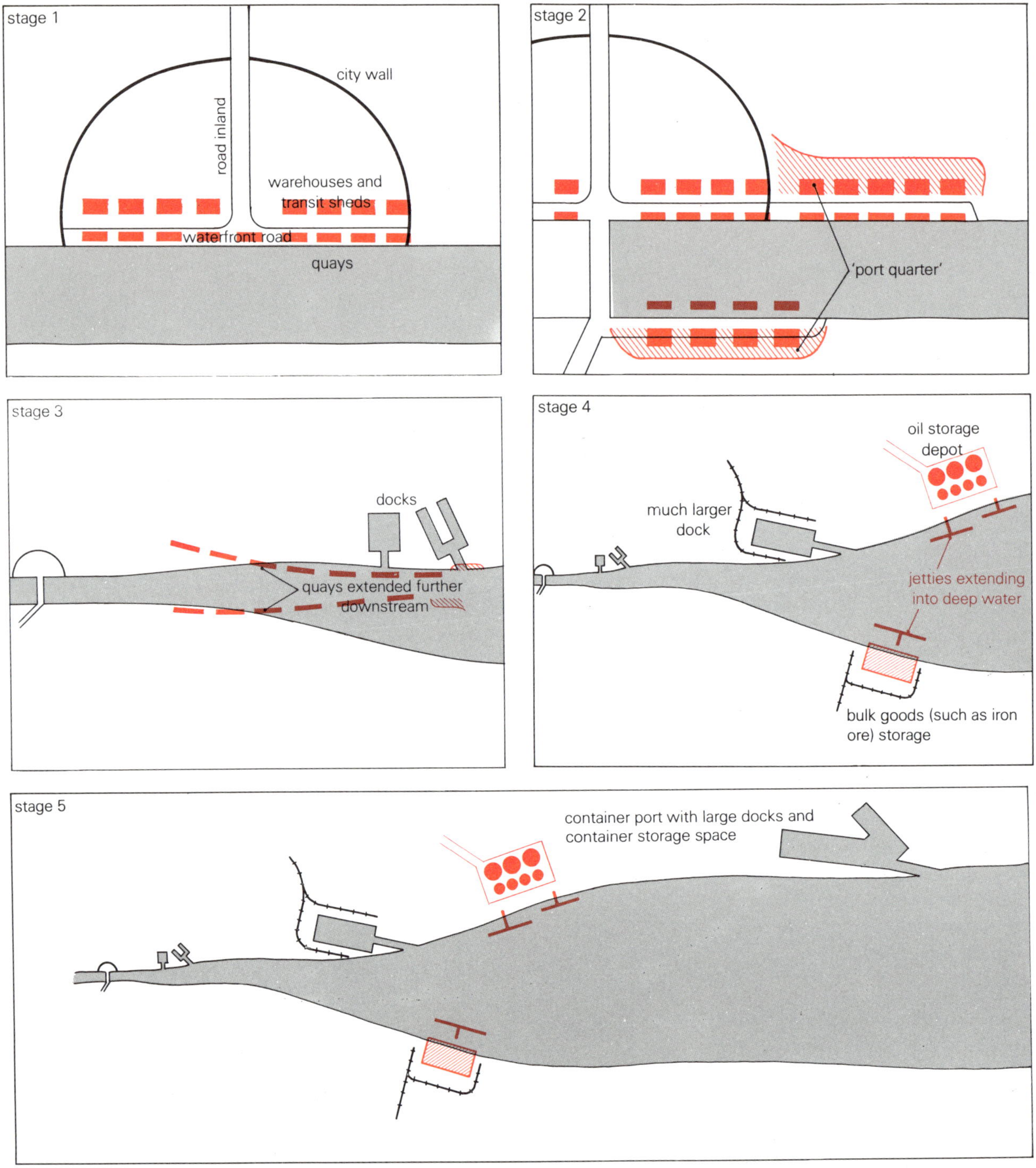

Figure 198 (a) Stage 1 (b) Stage 2 (c) Stage 3 (d) Stage 4 (e) Stage 5 Stages in the growth of a model port

A 'model' port

The container revolution is simply the latest in a series of developments which have been going on throughout port history. Most ports in the UK have followed a similar pattern. The history of an imaginary port might have gone something like this:

Stage 1 The port is founded when an area needs overseas trade. There is some sort of sheltered harbour, fairly deep water near the bank and roads which connect the waterfront with the producing and consuming area.

Stage 2 The original centre becomes crowded with shipping. Quays are constructed further downstream. These go beyond the edge of the original town, so that, around the new quays, a 'port quarter' develops.

Stage 3 Ships are becoming bigger and with the large tidal range, they tend to go aground at low tide. Therefore **docks** are built. Once again, these are further downstream.

4 (a) What is meant by 'tidal range'?

(b) What is the difference between a dock and a quay?

If you do not know the answers to these questions, look them up in an encyclopedia.

Stage 4 Ships are invented for carrying cargoes **in bulk**. This means that the goods do not have to be loaded in sacks, boxes or barrels, but can be poured, piped or lifted straight out of specially built holds. These new ships need deeper water than before, so another move downstream takes place. New, even bigger, berths are built. In some cases, particularly oil tankers, the berths may be on jetties which stick out into really deep water. A pipeline for unloading runs along these jetties.

5 What is meant by a 'berth'?

Stage 5 New inventions make rapid turnround possible, but they need more space. Further downstream again, a container port is built. This includes other new inventions, such as a roll-on/roll-off terminal, which will be explained later. Better ways of handling ordinary cargoes are also introduced.

Obviously, not all British ports are like this model. Where ports differ from the model, there are usually good reasons for the differences.

Modernising ports

Of the new inventions which fit in most neatly with the coming of containers, perhaps the most important is the roll-on/roll-off idea. This is not particularly new, because car ferries have been using similar ships for some time, but its use for large-scale freight, such as in figure 199, is new. Sometimes roll-on/roll-off

Figure 199 Ro-ro at Grimsby

Figure 200

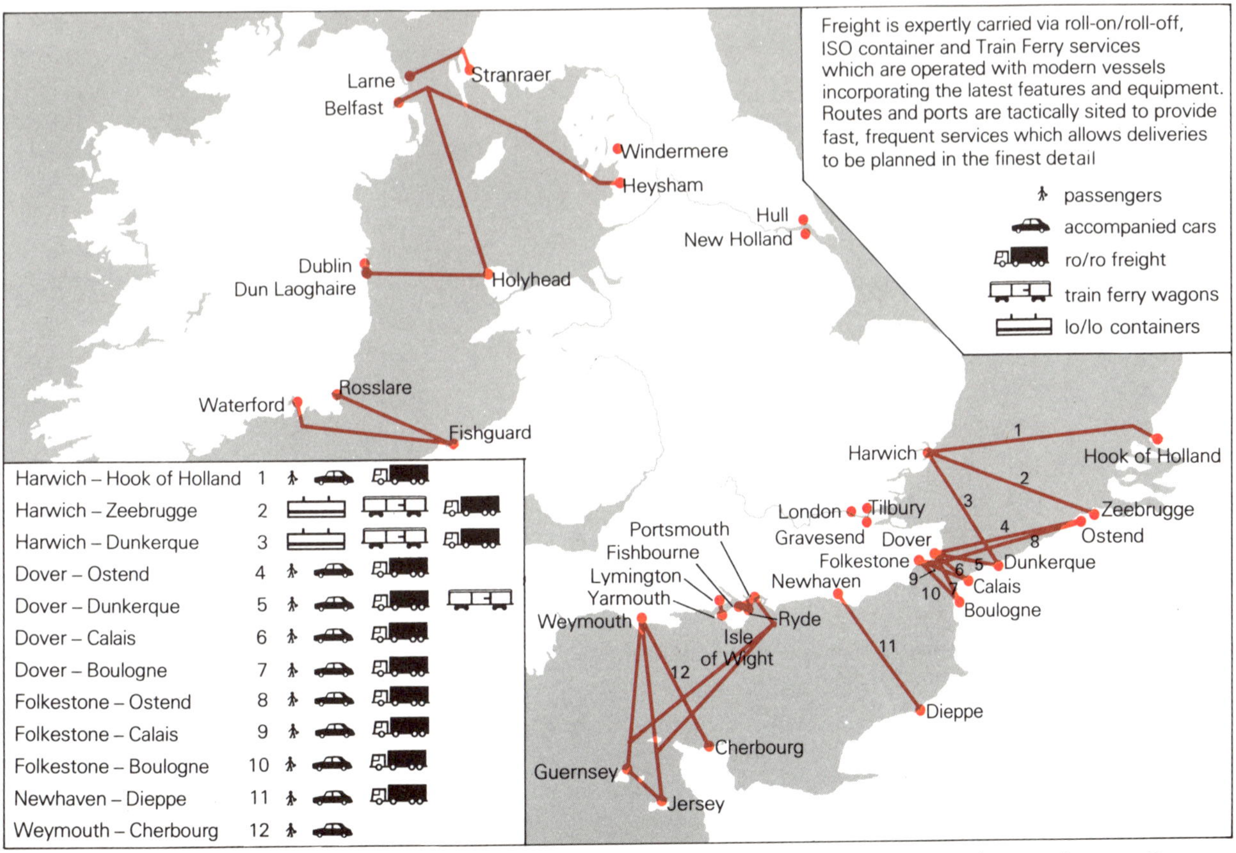

involves the whole lorry driving on to the ship and driving off at the other end. A variation of this is where the lorry drives on to the ship, unhooks its trailer, hooks on to another trailer which has been brought by the ship and drives off again. This method allows the lorry and its driver to continue working while the ship is at sea.

6 Which type of ro-ro is being used in figure 199?

A similar idea is 'Sealink'. Here the boats are owned and operated by British Rail, together with the railways of other countries. Attention is paid particularly to the ease with which passengers and goods can transfer from sea to land transport.

7 From figure 200, you can work out the details of the Sealink service:
(a) Across which seas does it operate?
(b) Can you think of any reasons why these routes are used and not others?
(c) Do you think ro-ro would be used as often on long distance routes?

Patterns of trade

One of your answers to question 7(b) must be that trade with Europe is becoming more important. The pattern of UK trade has been changing steadily ever since the Second World War. These changes have been greatly speeded up by membership of the EEC.

8 If you do not know what EEC stands for, find out. (See chapter 1 in *Spotlight on Europe*).

During the last century, as industries increased, a fairly simple pattern of trade grew up. Other countries supplied food and raw materials; in return, the UK exported manufactured goods.

	Thousand tonnes 1971
Foreign traffic	
Imports	
Total	203273
Foodstuffs	20282
Basic materials	34939
Manufactured goods	15202
Coal, etc.	4573
Petroleum	128278
Exports	
Total	48806
Foodstuffs	2179
Basic materials	8314
Manufactured goods	16802
Coal, etc.	3384
Petroleum	18127
Coastwise traffic	
Outwards	
Total	51190
Coal, etc.	9049
Petroleum	35529
Other	6612

Figure 201 UK seaborne trade

Let us now see if this pattern still continues.

9 (a) In figure 201, which is the largest import?
(b) Which is the largest export?
10 Petroleum is rather a special case, so look at the others.
(a) Of these, which is the largest import?
(b) Which is the largest export?

From looking at these figures, you may think that the traditional pattern of trade still goes on. But now look at figure 202.

Figure 202 is not quite the same as figure 201 because these figures are for the whole of the external trade of the UK, not just the seaborne trade. Also, the groups of commodities are very slightly different. Even so, the two sets of statistics are similar enough to be used for comparison.

	£ million 1971
Imports	
Total	9821·1
Food, beverages and tobacco	2175·8
Basic materials	1270·5
Mineral fuels and lubricants	1246·6
Manufactured goods	5003·2
Semi-manufactured	2571·8
Finished manufactured	2431·4
Exports	
Total	9181·4
Food, beverages and tobacco	587·5
Basic materials	278·4
Mineral fuels and lubricants	236·3
Manufactured goods:	
Total	7824·5
Metals	1004·5
Engineering products	4124·6
Textiles (excluding clothing)	425·6
Other	2269·8

Figure 202 Value of the external trade of the UK

11 Using figure 202, which is the biggest import by value?
12 What percentage of the total is this?
13 If you take out 'Mineral Fuels and Lubricants' (meaning mainly coal and oil) from the total, what percentage of the total is manufactured goods?
14 What is the largest export?
15 What percentage of the total is this?

Thus, by value, the traditional trade pattern no longer exists. Among both imports and exports, manufactured goods are by far the most important.

The *direction* of trade is also important. Where do our imports come from and our exports go to? Because the nineteenth-century pattern was exchanging manufactured goods for raw materials and

	1964		1973	
	value £	%	value £	%
Total	10261·1	100	28309·2	100
Sterling area outside Western Europe, plus Canada	3610·0	35·2	6166·8	28·8
N. America: total	1732·0	16·9	4293·6	15·2
Canada	651·9	6·4	1149·6	4·1
US	1080·1	10·5	3144·0	11·1
Europe: total	3953·3	38·5	14144·4	50·0
EEC: total	2567·5	25·0	9226·8	32·6
EFTA: total	1093·3	10·7	4117·2	14·5
Rest of World	1617·7	15·8	4854·0	17·1

Figure 203 *UK overseas trading areas*

food, it is often thought that most of our trade was with the Commonwealth. This has never been the case, except for a short period after the Second World War. Naturally, from the eighteenth century onwards, two-way exchanges with the Commonwealth have been considerable, but throughout most of our history, most trade has been with Europe.

In figure 203, you can see the recent position. The year 1973 is important, because it was the first year of British membership of the EEC. In these statistics, the clumsy heading 'Sterling Area Outside Western Europe plus Canada' more or less equals the Commonwealth, and recent ex-Commonwealth countries. The heading 'Western Europe' means non-communist Europe.

16 In figure 203 look at the two areas just described. What percentage of the UK trade did each of them have in 1964?

17 Now look at the same two areas in 1973. What were the percentages in that year?

18 In your own words, describe the *trends* in UK trade which are suggested by the two comparisons you have just made.

The EEC is, of course, *part* of Western Europe, but even if you take it on its own, you will see that it was second after the Commonwealth in 1964 and some way ahead in 1973.

As far as trade with individual countries is concerned, the USA is far ahead of all the others, with West Germany in second place. It is interesting that both of these are highly industrialised countries.

19 The figures in figure 203 are by value. If figures for weight were used, the Persian Gulf area would be among the leading traders with the UK. Why do you think this is?

The ports' 'league table'

On pages 132 to 135 of this chapter, the nature of ports was explained. Figure 204 gives a 'league table' of ports, but before you look at it, see if you can work out what the positions are likely to be using the knowledge you already have.

20 Are the ports which developed during the Industrial Revolution still likely to be the most important?

21 Is the position of those ports closest to Europe likely to have changed?

22 Are the ports in the South East

		Percentage of total trade by value
Year: 1938		
1	London	38
2	Liverpool	22
3	Hull	6
4	Southampton	5
5	Manchester	4
6	Glasgow & Greenock	4
7	Bristol	2
8	Newcastle	2
9	Harwich	2
10	Swansea	1
Year: 1960		
1	London	36
2	Liverpool	23
3	Hull	6
4	Manchester	5
5	Southampton	4
6	Glasgow & Greenock	4
7	Bristol	3
8	Harwich	2
9	Swansea	2
10	Dover	2
Year: 1971		
1	London	22
2	Liverpool	13
3	Heathrow Airport	11
4	Southampton	6
5	Harwich	5
6	Hull	5
7	Felixstowe	4
8	Dover	4
9	Glasgow & Greenock	3
10	Manchester	3

NOTE: (a) The percentages do not add up to 100 because they are percentages of the total trade of *all* the ports of the UK. (b) Although some percentages are the same, the rank order of the ports is correct because the percentages have been rounded off to the nearest whole number.

Figure 204 *Ports' league table*

Figure 205

likely to have developed more than others?

23 Are there likely to be any 'specialist' ports, which import only one commodity in bulk?

24 Draw a divided circle graph (see page 70) for each of the years. From this you can see at a glance the order of importance of the ports and the size of 'the slice of the pie' which each port has.

25 The same two ports have dominated the top ten for many years, but in 1971 they had a new competitor in third place. What was the name of this port?

In the table the position of individual ports is interesting. Some have maintained their position, some have come up and some have slipped.

26 What has happened to the league table position of Harwich, Felixstowe and Dover over the period covered by figure 204?

27 In what part of the country are these three ports located?

28 What has happened to the positions of Manchester, Glasgow and Greenock, and Newcastle?

29 In which part of the country are they situated?

Air transport

One of the most interesting facts to emerge from figure 204 is that Heathrow Airport is now the UK's third port by value of goods. This may be surprising, but simply shows two bits of information about air transport. The first is that since the 1950s, air transport has increased dramatically in importance all over the world. The second is that aircraft are ideal for carrying very high value

Location	Airports	Passengers (000's)	Per cent of total	Change from 1970
Airports				
Birmingham	Elmdon	553	1·5	+40·4%
Glasgow	Abbotsinch	415	1·1	+19·3%
Liverpool	Speke	158	0·4	+30·6%
London	Gatwick	4146	11·3	+27·6%
London	Heathrow	13455	36·5	+ 6·8%
Luton	Luton	2668	7·2	+38·7%
Manchester	Ringway	1419	3·9	+36·6%
Prestwick	Prestwick	309	0·8	− 0·6%
Southend	Rochford	441	1·2	+ 8·6%
Stansted	Stansted	474	1·3	—
Others		1087	2·9	+21·0%
Total		25125	68·1	+16·1%
Seaports				
Dover		5157	14·0	+ 2·5%
Fishguard		353	1·0	− 1·2%
Folkestone		903	2·4	− 6·7%
Harwich		1284	3·5	+ 4·2%
Holyhead		298	0·8	−33·3%
Hull		186	0·5	+ 7·5%
Liverpool		438	1·2	—
London		119	0·3	+36·7%
Newhaven		504	1·4	+ 2·0%
Southampton		1015	2·8	+ 0·6%
Tyne		132	0·4	− 0·8%
Others		1348	3·7	+19·7%
Total		11737	31·8	+ 2·1%
Total airports and seaports		36862	100·0	+11·2%

Figure 206 International passenger traffic, 1971. Source: British Airports Authority

Figure 207 Interior of a Boeing 747 Jumbo jet

cargoes which do not weigh very much, such as jewellery or currency. On the other hand, aeroplanes are not much use for very bulky goods. You may feel that taking value alone is not very fair. The UK would soon grind to a halt if the bulk goods like petroleum, iron ore and grain stopped coming in. So the seaports which handle these must be considered vital. Heathrow comes nowhere in a league table of weight of goods handled.

Figure 205 shows that aircraft do carry a varied cargo, even if not of bulky goods.

30 Make a list of ten things which you think might be carried by air rather than by sea.

Despite the increasing importance of cargo traffic, the impact of air transport has been greatest on passenger trade.

Look at figure 206.

31 What percentage of the UK's international passenger traffic is handled by airports?

32 Draw a graph of the top ten passenger ports in order to compare them.

33 Which is the leading seaport for passengers?

One of Dover's advantages is that the sea routes to Europe are very short. Even so, nearly twice as many people travel to Europe by air as by sea. In 1971, 5 205 000 people travelled by air compared with only 246 000 by sea in journeys to other parts of the world.

Moreover, the number of air travellers can be expected to increase. You may have been on a 'package tour' holiday by air. This type of air transport is rapidly increasing, in addition to the increase in scheduled (timetabled) flights.

One of the reasons for the increase in passenger travel by air is a new invention. Technological change is, once again, altering transport. In this case it is the arrival of the wide-bodied jet.

34 How many seats can you see side by side in figure 207?

In fact 747s, and the shorter range Tri-Star and DC10, can seat some 350 people, well over twice as many as the previous 'big jets'.

Environmental problems of air traffic

Concorde is another form of 'technological change'. This aircraft could change the pattern of transport yet again, but even before it went into service there were many complaints about it. People in many countries have tried to have it banned from flying in their airspace. One of the reasons people object to it is the sonic boom, a deafening double bang. This will mostly only happen over the sea or deserts. But there are other objections to Concorde which are shared by other aircraft. Imagine being underneath the plane in figure 208.

Figure 208

Maidenhead
M4
Slough
40
West Drayton
Ealing
M4
Brentford
Chiswick
Windsor
Hounslow
60
50
Staines
40
Twickenham
Richmond
Egham
Sunbury
40
River Thames
Walton
school
hospital
-40- Noise and number index (NNI)
contour line
0
5 km

Figure 209 (opposite) *Noise and Number index map for Heathrow*

Anyone who has been underneath an aeroplane taking off will be well aware of the ear-splitting noise. Lengthy exposure to this kind of noise can cause some ill-health. Aircraft noise does not only occur near the airport. For a long way at either end of the runway, there is an area in which normal conversation is hindered and in which hearing the radio or television is difficult.

The noise area round an airport can be shown by a kind of 'contour map'. This attempts to show the levels of noise, combined with the number of times during the day that the noise happens. It is called a 'Noise and Number Index' map. A Noise and Number Index of over 50 causes severe disturbance to people.

35 (a) In figure 209, how many hospitals are there inside the 50 NNI line?
(b) How many schools?
In addition, several hundred thousand people live within the line.

Look at figure 210. It is hardly surprising that there are campaigns to cut down aircraft noise. Yet the solutions are not all that easy. It is not possible to change many of the routes, simply because the planes have to land and take off along the runways. Only after take-off can some routes be varied. Even on these it is difficult, because aircraft navigate by radio beacons, and these have to be moved if the route is changed. Even if you do this, you are probably relieving some people of the misery, only to pass it on to others. The only real cure is probably another technological change. Quieter engines are being developed (though not in Concorde) but hope for the distant future lies in the invention of vertical take-off planes, which do not need a long run-in over built-up areas.

Another reason why change is not easy is the importance of Heathrow. It is by far the busiest airport in the UK and, for international flights, is the busiest in the world. Yet this same traffic causes problems other than noise. The air is so full that air traffic control is an immense strain. Even with the latest computer and radar equipment, individual officers need a rest after two hours or so working at a busy time.

At peak periods, aircraft are landing at something like every ninety seconds, others are taking off at the same time and there is constant coming and going all around the London area.

Not only is the air crowded, but the airport itself is full. The runways are in almost constant daytime use. The aircraft ground movement and parking areas are overcrowded and the terminals are having to cope with too many people.

Perhaps one of the worst aspects of Heathrow is the way in which the terminal buildings, car parks, etc., are all in the centre. This seemed a very good idea when it was built, but no one could have forecast the great increase in traffic. Now the tunnel which links the central area with the M4 and A4 roads is also overcrowded. If one of the lanes in the tunnel has to be closed down for any reason, great traffic jams build up and flights are missed.

36 Read in figure 211 what some passengers had to say about one such hold-up.

London's third airport – Will it ever be built?

What can be done about Heathrow's overcrowding? The obvious sug-

Figure 210

Figure 211

Honeymooners hit in airport chaos

ACTOR Richard Attenborough had a few choice words about the chaos he surveyed at Heathrow Airport yesterday.

It's an absolute bloody disgrace," he bellowed. "They ought to have their heads examined."

Thousands of other air travellers were equally furious at being caught up in one of the most massive traffic jams Heathrow has known.

And all because of a "Road Up" sign on an airport road leading to the long-haul terminal building.

Workmen had the road up as a part of a job on the new

Heathrow 'disgrace' of worst-ever jam

underground extension of the Piccadilly Line.

The result was tremendous queues at the solitary tunnel which arriving traffic has to pass through.

At one stage not a single car moved for half an hour and the jam caused a traffic tail-back on to the M4 motorway.

Disastrous

Things got even worse as the day went on. It took a United Arab Airlines representative 90 minutes to get from the tunnel entrance to the airport central area.

Thirty-four-year-old David Jackson from Lincoln, due to catch the 13.05 flight to Zurich, arrived at 14.30. Because of the jam the drive from London took him two and a half hours—compared with less than two hours to fly to Zurich.

"This," said Mr. Jackson, "must be the most disastrous day in Heathrow's history. The whole situation is beyond description."

Hundreds of passengers were delayed in getting away. Many more missed their flights altogether including a Palma-bound honeymoon couple.

Families due to fly off on late summer holidays walked round the crowded terminal building in a daze. Some women were in tears.

Eventualy, to ease the congestion, British Airports Authority opened the airport's cargo tunnel. By then many passengers had found themselves stranded. Richard Attenborough was luckier. He just managed to catch his flight to Nice. John Robertson, from Wimbledon, had no such luck. He missed a flight to Scotland by 20 minutes.

"I just can't think of words foul enough to describe this place," Mr. Robertson stormed. "The whole operation is just an abominable mess."

Criticised

An Air France official was equally critical of B.A.A. "Why the heck they decided to do all this road work at the peak period of summer I don't know," he said.

"The irony is that if a plane is delayed because of late passengers the airline has to pay B.A.A. extra fees. They cause a mess, then make a profit. It's totally disgusting."

But B.A.A. claimed most of the trouble was caused by the large number of people at the airport just greeting passengers or seeing them off.

gestion is to build a third airport, in addition to Heathrow and Gatwick, for London.

Having decided to do this, the government of the day chose to expand Stansted airport, which had relatively little traffic. Strong opposition eventually succeeded in squashing this scheme. After that, the Roskill Commission tried to decide the best location for a new airport. The factors of location for an airport can be analysed in the same way as for an industry.

37 The airport should be **accessible** to the maximum number of people. In order to achieve this in the UK, where, approximately, would you locate a new airport?

38 In the area you have chosen for question 37, where would you locate the airport so that it was *also* near a motorway and main line railway?

39 Unlike Heathrow, an airport is best situated where the flight paths do not cross heavily populated areas. Is this possible if the airport is to be accessible to the maximum number of people?

Another location factor is a large area of firm flat land, but this is often the best farming land, so some sort of conflict is likely there too.

The cost of building an airport is enormous. It helps, therefore, if as many as possible of the factors suggested above can be met without extra building. This makes the project more 'cost effective'. That is to say, the money you put into it will be less compared with the benefits you will get from it. Using this approach of 'cost/benefit analysis', the Roskill Commission suggested Cublington as the site for London's third airport.

Figure 212 Sites considered by the Roskill Commission as possible locations for London's third airport

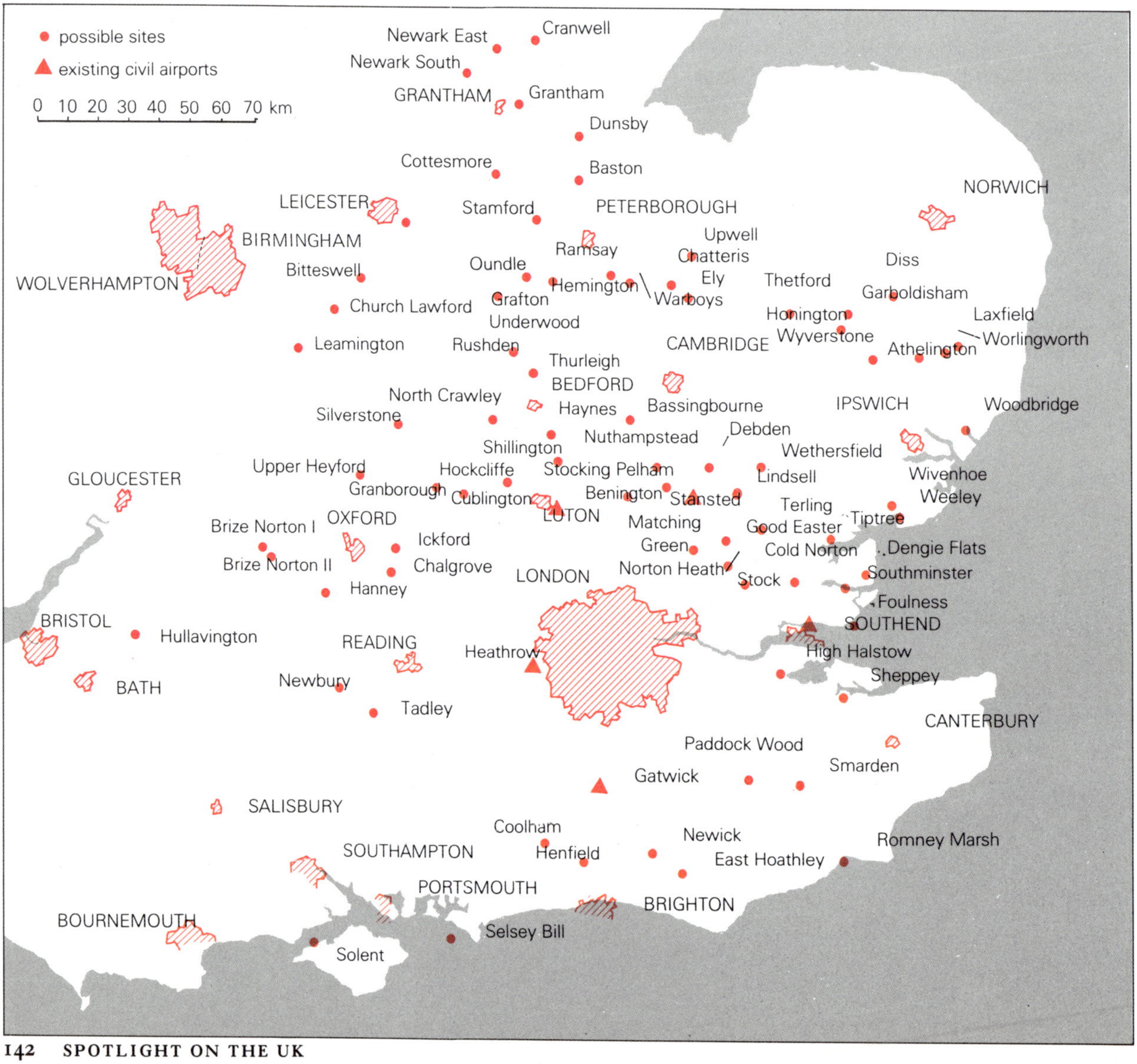

40 Write down why you think Cublington was chosen by the Roskill Commission.

Eventually, the government decided on a site at Foulness, on Maplin Sands. This was going against the advice of the Roskill Commission. On a purely cost/benefit basis, the Maplin site would be very much more expensive.

Among the benefits would be the ability to construct an entirely new seaport as yet another extension of the overcrowded London Docks. But the main advantage was the fact that many of the flight paths would be over water, and would therefore not disrupt the lives of too many people.

41 You will see from figure 213 that not everyone agreed that there would be very little disturbance from aircraft noise. Name some of the areas which would have been affected.

There were many complaints about the choice of Maplin. Nobody wants an airport on their doorstep, and many reasons were given for not having it there. But the strongest attacks came from those who thought that a new airport was not needed in any case. The wide-bodied jets can carry so many people that fewer flights are required. These aircraft are also quieter than the earlier ones. More progress is being made on silencing the engines of existing types. There is the possibility, at some future time, of vertical take-off planes, which would hardly affect the surrounding area at all. Will there ever be a 'third airport'?

The Channel Tunnel

Now suppose that the Channel Tunnel were to be built. In this case,

Figure 213 Maplin Airport as described by the Sunday Times

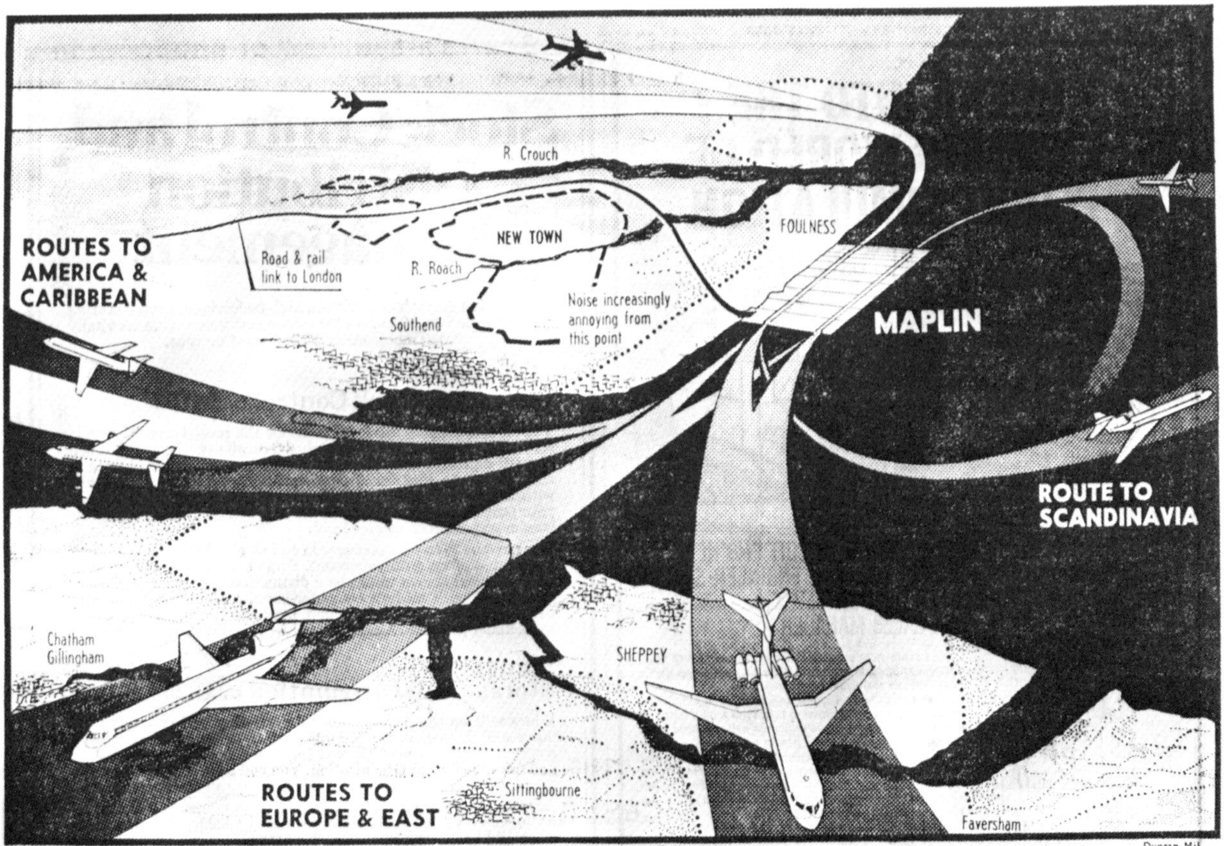

Maplin's air routes

Contrary to common belief, most air movements to and from Maplin will be audible over land (the dotted line is the outer "contour" of serious noise nuisance). The drawing and map (right) are adapted from an official but unpublished operational map which takes into account the realities of air traffic control—showing heavy traffic landing over Essex and taking off over Kent.

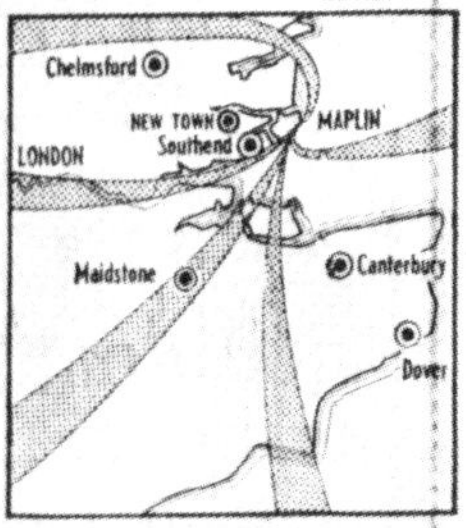

many cities in the western part of Europe would be easily accessible by rail. If high speed links were to be built, as shown in figure 214, the time from city centre to city centre would be as quick as by air, and there would be fewer delays caused by weather.

The Maplin project has been cancelled for a mixture of the reasons given above. Whether the Channel Tunnel will ever be built remains unsure, but the government is no longer giving its backing to it. The benefits just mentioned would follow from the construction of the tunnel, but can the country afford the massive cost involved? The tunnel, too, would cause many environmental problems.

Space has just been used to study a transport project which has been abandoned and one which may never take place. Why? It is because they show very clearly the pressures on the UK transport system. Transport is the lifeblood of the nation. In the modern world, no country can exist in isolation. In addition, quite apart from the actual transporting of things, the business itself provides employment for hundreds of thousands of workers. Heathrow alone employs 50000 and British Rail, in all its varied activities, employs about 250000.

42 Look through an AA book and find a few towns of the same size as Heathrow's labour force.

But although transport is so important, we must not lose sight of values other than simple economics. It is no good being a rich nation if the country is unfit to live in. If you really understand the issues involved, you can help to ensure that the UK is fit to live in.

PROJECTS

1 See if you can discover any information about some of the ports that are on their way up in the league table (page 137), such as Harwich, Felixstowe, Heathrow or Southampton. If you can find sufficient material (the Port Authorities might help you), you could split into groups and each group prepare a display on one port.

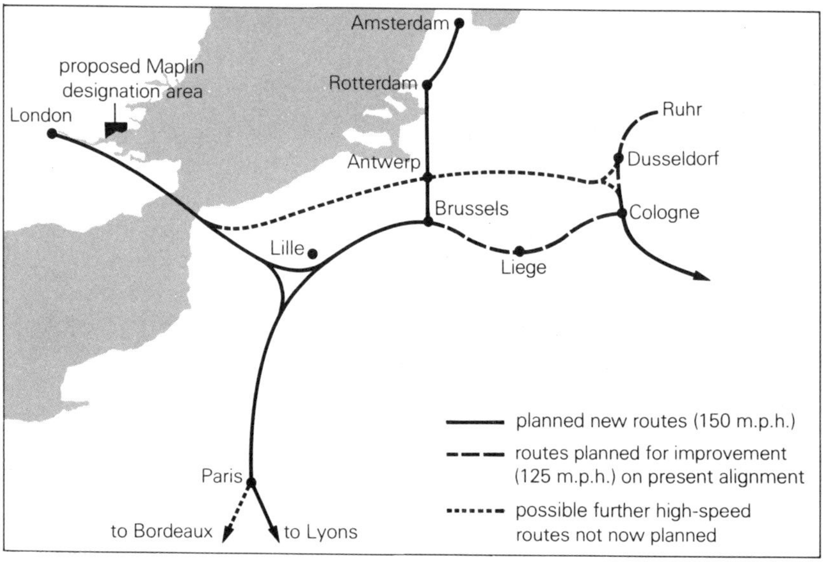

Figure 214 Proposed high-speed railways in Europe

2 Edinburgh needs a new airport. The existing one, Turnhouse, is too small and badly designed. Decide on the best location for the new airport. Use 1:50000 sheets 64, 65, 66. Some facts to help you are as follows:

(a) There is to be a single runway 2600 metres (8500 ft) long.
(b) The runway is to run as nearly as possible from south-west to north-east.
(c) The noise disturbance area is an oval, about 20 km (12 miles) long and 3 km (just under 2 miles) wide. This is not as big as Heathrow's because there are many fewer aircraft taking off and landing.

In selecting a site, it will help to draw an oval to scale, with the runway marked in the centre, on a piece of tracing paper and then move this around over the map.

Using a 1:50000 map, the oval will be about 40 cm long and 6 cm wide. The runway will be 5·2 cm long.

Using a one-inch map, the oval will be about 31 cm long and 5 cm wide. The runway will be 4·2 cm long.

Some opinions which may help are:

(a) The inhabitants of Crammond and Newbridge do not want the airport to be next to Turnhouse because the new flight paths would be over their residential areas.
(b) The business people of Edinburgh think a new airport fit for Scotland's capital city is vitally needed.
(c) Other people think it is a waste for both Glasgow and Edinburgh to have airports. There should be a new international airport to be shared by the whole of central Scotland. (An international airport would need a runway 3700 metres (12000 ft) long. The proposed 2600 metre runway is suitable for only short and medium haul aircraft. If you wanted to go to America, you would have to go to Prestwick, Glasgow's airport).
(d) Many people think that Edinburgh will need to expand to the west in the future.
(e) The County Council think that the Edinburgh airport is very necessary.

12 Fish and chips

An early farming pattern

In, say, the late eighteenth or early nineteenth century, when all transport of goods was by animals or humans, the decision of a farmer about what crops to grow had to take this factor into account.

If all the land was suitable for arable farming (that is, for growing crops), the decisions might be made as follows:

(a) on a farm close to a town, high-value goods would be grown, such as fruit or vegetables.
(b) A little further out, milk could be produced for the town dwellers.
(c) Further away still, cereals could be grown.
(d) Beyond this area, farms might produce cattle and sheep for meat and wool.

The result of such an arrangement of farms around a town would be a series of 'zones', each producing something different (figure 215).

Figure 215 A possible arrangement of farming zones around a town of the late eighteenth century

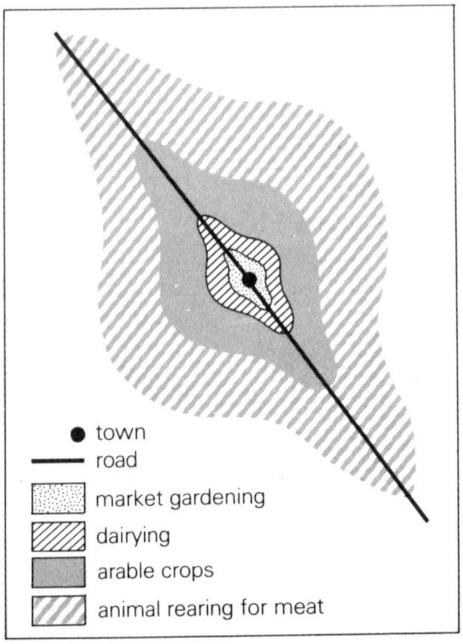

1 Suggest reasons why this arrangement was likely at that time.

Location in farming

The farmer making the decision of what to grow, is very similar to the industrialist deciding where to locate his factory (page 16). In both cases, transport costs are important. For the farmer, the location is fixed and he must decide what to grow depending on how much it will cost to take it to market. The industrialist knows what his product is, and must decide where to put his factory to reduce transport costs.

As we saw in chapters 2 and 3, there is a great deal more to location than the simple idea given above. In the present age of high-speed transport, nowhere in the UK is very far from a large town, or a conurbation. So the model of land use zones in figure 215 is no longer true, but, as with other models, the idea is a useful starting point for looking at the actual way the land is used.

But is it really out of date? What about the pattern on any one farm? Imagine you are a farmer whose land is suitable for arable farming.

(a) You could make a profit from vegetables, but these need a great deal of time and attention spent on them;
(b) You can make a good profit producing milk, but this too demands a great deal of labour, and a lot of money to be spent on equipment.
(c) You can make less profit on cereals, but on the other hand, they do not require all that much labour, except at harvest time. Also, using machinery, you can grow large quantities.

Suppose you decide to produce all three. This is quite a good idea as it spreads the risk. The risk may be from the weather. Storms at harvest time could ruin the cereal crop but leave the vegetables intact. The risk could be economic. The price of milk might fall but that of cereals remain high. Farmers always have to take risks with the weather and with prices in this sort of way.

2 You have decided to grow all three crops.
(a) Which would you plant nearest to the farmhouse? Write down the reasons for your answer.
(b) Which would you plant furthest away? Again, give reasons.

In this example, the model may still have some value, but there are many other things which can upset it. What about the type of soil and the climate? The UK is a small country but there are great variations from one part to another. Therefore the type of farming changes from one part of the country to another – but only to a certain extent. If it is worthwhile to grow a particular crop – if the profit to be made is great enough – then it is possible to grow it on unsuitable soil. It comes back to economics. If the soil is not the best kind for the crop, then it will cost money to grow that product. For example, clay soil might have to be artificially drained, or chalky soil may need clay added. These two processes would both be expensive, but if the income from the chosen crop was great enough, it would be worth it.

Even so, the differences in climate from one part of the country to another *are* great enough to produce differences in farming type. There is no crop in the UK which is so valuable as to make farmers grow it everywhere at whatever cost.

Compare figures 216 and 54 and see how similar they are.

3 Which relief features are found in the areas of Chalk and of Jurassic limestone?

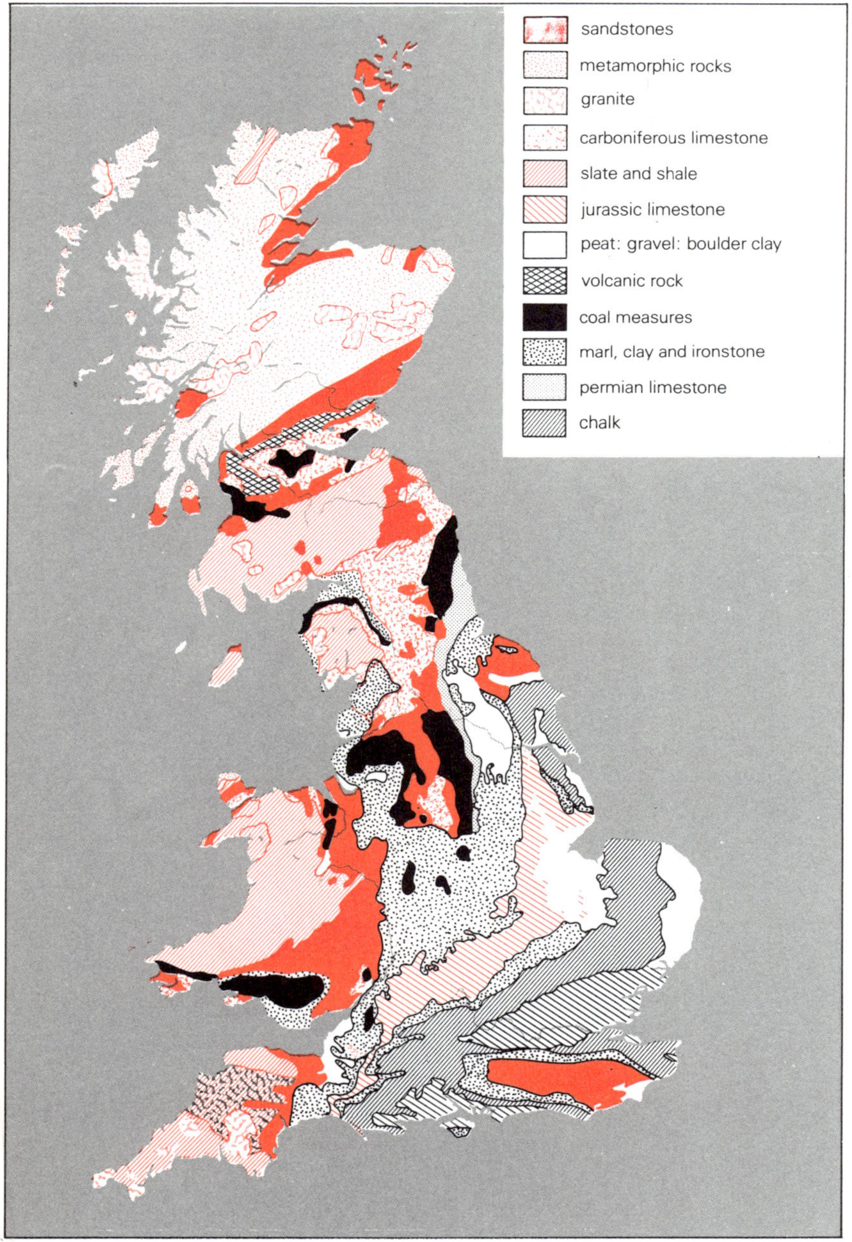
sandstones
metamorphic rocks
granite
carboniferous limestone
slate and shale
jurassic limestone
peat: gravel: boulder clay
volcanic rock
coal measures
marl, clay and ironstone
permian limestone
chalk

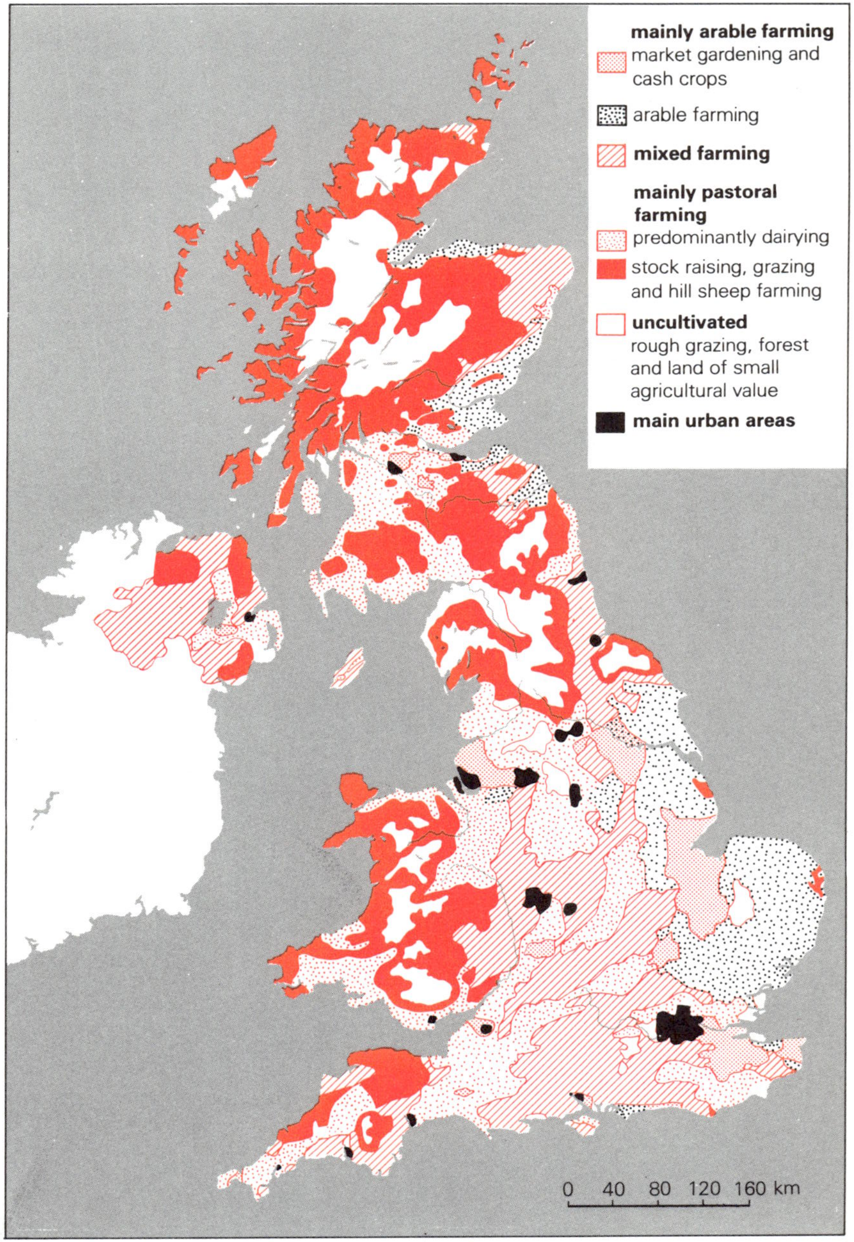
mainly arable farming
market gardening and cash crops
arable farming
mixed farming
mainly pastoral farming
predominantly dairying
stock raising, grazing and hill sheep farming
uncultivated
rough grazing, forest and land of small agricultural value
main urban areas
0 40 80 120 160 km

Figure 216 (opposite, left)   Geology of the UK
Figure 217 (opposite, right)   Types of farming in the UK

4 Are the main areas of clay highland or lowland?

5 What are the main types of rocks found in the highland areas of the UK? (The Scottish Highlands and Southern Uplands; the Lake District; North and Central Wales; Dartmoor and Exmoor in the South West Peninsula.)

Look at the work you did for the project at the end of chapter 5, where you found out about the climate of the UK, and the way it might affect your holiday choice.

6 Which relief regions of the UK are the wettest?

7 Which relief regions have the coldest summers?

Types of farming

8 Compare figures 216 and 217. Can you see any ways in which they are similar to each other?

9 What types of farming are found in the main highland areas?

10 In the east, from north of the Humber almost to the Thames estuary there are only two main types of farming. What are these? Figures 218 and 219 give a clue. One of the pictures is of the Fens. The other shows part of Kent where one of the two kinds of farming is also found.

11 In what kind of geology and relief regions is dairy farming mainly found? You will see that it is very widespread but two types of physical region are particularly important. To help you with this question, concentrate on two areas.

(a) What kind of geology and relief is found in the dairying area to the south of Liverpool?

(b) What is the relief like in the dairying area of the Scottish Southern Uplands, to the south of Glasgow?

Figure 218 (above)
Figure 219 (below)

Look again at figure 217 and you will see that mixed farming is to be found in almost all parts of the UK. On many of these farms, dairying is an important part, so that it is found over an even greater area than might have been thought from the map. In fact, it is found on almost all types of land, apart from the very worst mountain areas and the very best arable land. But there can be many kinds of mixed farm. One sort might be mainly arable with a few cows to manure the fields in a crop rotation and provide some safety if the crops are high risk types.

12 What is meant by 'crop rotation'?

13 What do you think 'high risk' crops are?

At the other end of the mixed farm scale may be a dairy farmer who grows a few crops to provide additional food for the cattle. In between these two extremes of mixed farming is a whole range of types.

Processes changing agriculture

The types of farming map (figure 217)

	England & Wales (% of total area)			
	1900	1925	1950	1965
Agriculture	83·6	82·9	80·6	78·6
Forest & woodland	5·1	5·1	6·4	7·5
Urban land	5·4	6·2	9·7	11·5
Land unaccounted for	5·9	5·8	3·3	2·4

Figure 220

is a good guide, but it is so general that it conceals a great deal of detail. Although on the whole there is a difference in farming between Highland Britain and Lowland Britain, some types are so widespread that they can be found almost anywhere. Because of this it is, as usual, important to know what the processes affecting UK farming are, rather than to know the facts as they exist at any one time.

The first, and probably the most important, influence on farming is government policy. Before the Second World War, over two thirds of our food was imported. The war made us grow more at home and in the early 1970s we produced more than half of our own food. Whether this process continues will depend on how important the government thinks it is, and, therefore, how much support they give to farming.

Another process of change, closely connected with the amount of food we produce, is the gradual loss of land for farming.

Look at figure 220.

14 What was the percentage of the total area of England and Wales used for agriculture in 1900?

15 What was the percentage in 1965?

16 Which types of land use have increased their percentages?

The information for England and Wales, rather than the whole UK has been used because the huge area of rough grazing in the Scottish Highlands is included as agricultural land, which distorts the picture.

Farmers, naturally, do not want to lose their land, and they point out how important it is for producing our food. On the other hand, the population of the whole British Isles

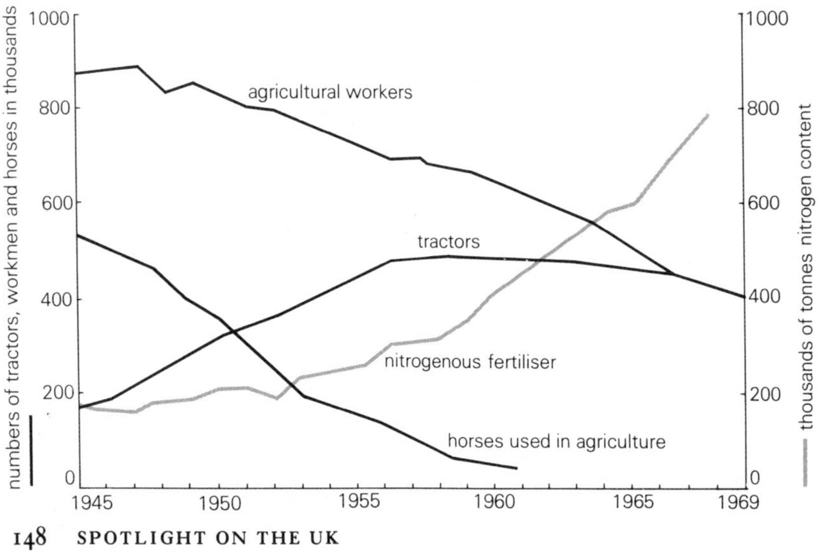

Figure 221 (above)
Figure 222 (below)

(that is, including the present Republic of Eire as well as the UK) was 41.5 million in 1901. In 1971 there were 55·6 million people in the UK alone. They all have to live somewhere.

17 What do you think the farmer in figure 221 is pointing to?

Land is not the only thing that has declined in the agriculture industry.

18 If you are leaving school soon, are you thinking of taking a job in agriculture? Do you know whether any of your friends are likely to go into this work?

If you live in a town it is unlikely that you know anyone who wishes to work on a farm, but even in the countryside there will not be many. There are many reasons why the agricultural labour force is getting smaller. The wages are poor compared with industry, and machinery is being invented to enable fewer men to produce more. In remote areas, people are leaving for social reasons. (Remember rural depopulation in chapter 9.)

19 Look at figure 222. Approximately how many agricultural workers were there in 1945? How many were there in 1970?

The graph in figure 222 shows only agricultural workers. That is to say, farmers and farmers' wives are not included. Taken together, workers and farmers still only add up to about 700 000 which is a mere 3 per cent of the total labour force of the country.

So far, it would appear that farming is declining, yet this is not the case. Despite the reduced area and the reduced manpower, productivity has steadily increased.

Figure 223 (a) From this . . . (b) through this . . . (c) to this . . . (d) and this. Three ways of harvesting potatoes, and a combine grain harvester

Figure 224 A modern milking parlour

Modern farming methods

Advances in machinery design, like those shown in figure 223, mean that more work can be done by fewer people. Scientific breeding of plants and animals produces bigger and better stocks. In addition, new ways of fighting disease and pests are being developed. Equipment in the farm buildings is also continually being improved.

In the milking parlour shown in figure 224, the cows are fed automatically, the dairyman does not have to bend and the milk never sees the inside of a bucket. It is pumped direct from the milking equipment to individual containers, which show how much each cow is producing. From there, it goes to a large refrigerated storage tank. A tanker lorry collects the milk from the farm.

20 Can you think of two sorts of places to which the milk might be taken after leaving the farm?

Look at the background of figure 225.

21 Do you have a greenhouse? What is the purpose of a glasshouse?

In the foreground of figure 225 are strawberries. Glasshouses and strawberries. All very ordinary you may think. But what you cannot see is the underground cable which is heating the strawberries and enabling them to ripen several weeks earlier than they would otherwise.

Most people are aware of the ancient practice of irrigation in semi-desert countries, but do you know of its use in the UK? Figure 226 shows two of many different modern irrigation machines.

Irrigation illustrates another example of conflict over resources in the UK. In Highland Britain, there is too much water, so there is no problem in that region. In the South East there is too little. And, of course, it is in the South East that most of the users live. Industry uses vast amounts and ordinary households use quite a lot. By comparison, farmers do not use all that much but, naturally, they wish to use it most when it is in shortest supply.

Figure 225

Who should have the water in the South East? If farmers were allowed to use more, they would be able to produce more food. Or should it go to industry which would allow us to buy more from abroad? Is it not

Figure 226 (right and centre)

possible to supply more water in the South East by transferring it from the upland areas? This is too expensive. How about storing more water? This is possible, but uses up valuable farmland. New ways of storing water and of purifying it are being examined, but at present these are for the future. Until they are developed, the arguments over who should have the water will continue.

Another process of change concerns the size of fields. Modern equipment is very expensive and it costs more to use it in small spaces. So farmers often try to extend their land by buying fields, or complete farms, when their neighbours retire. Frequently they enlarge the size of fields by digging up the hedges.

22 How far do you think it is to the nearest field boundary in figure 227?

23 Can you think of any problems which might be caused by enlarging fields like this? (Imagine the wind blowing, unbroken by hedges or trees, over that cultivated soil.)

Specialisation

Because they have to buy and maintain expensive equipment, farmers are increasingly finding that they cannot afford the installations needed for more than one type of farming. Most farming is still mixed, but the trend is towards specialisation.

Look at figure 228.

24 Norfolk has long been one of the chief counties of the UK for arable farming.

(a) What has happened to the percentage of arable land in Norfolk between 1950 and 1969?

(b) Do you think this shows anything about specialisation?

Figure 227 Cultivating in the fenland

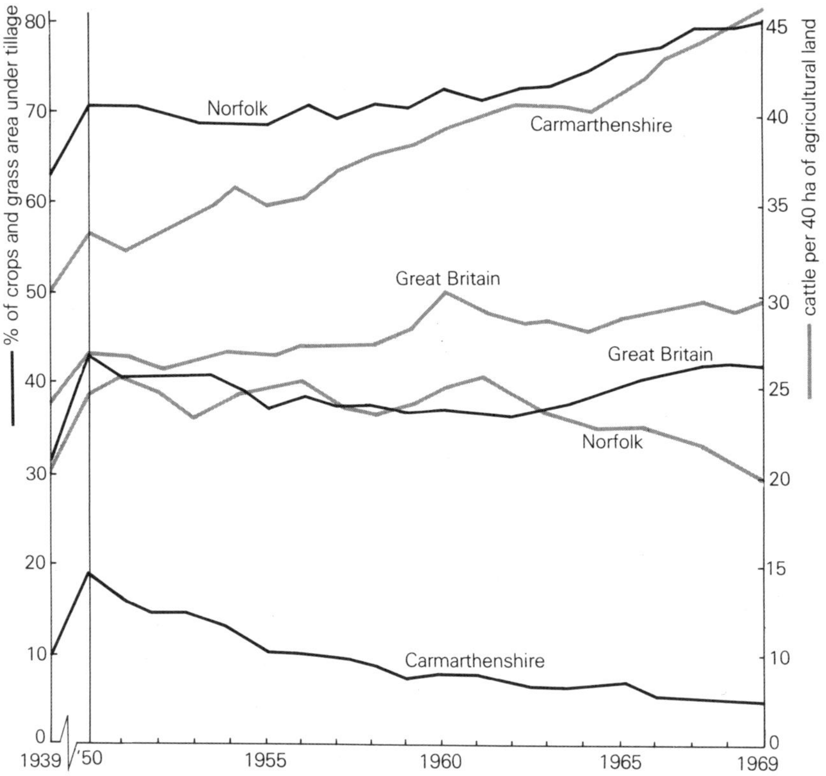

Figure 228

25 What has happened to the percentage of arable land in Carmarthen over the same period?

26 Carmarthen is one of the main counties for cattle rearing.

(a) What happened to the number of cattle per 40 ha. between 1950 and 1969?

(b) Do you think this shows anything about specialisation?

27 What happened to the number of cows in Norfolk during the same period?

Farming and the countryside

The tendency towards greater use of machinery, larger farms and fields and more specialisation, is resulting in farms becoming much more like industries. The modern farmer must not only know how to farm but must be a good business manager as well.

This 'industrialisation' of farming is leading to more argument. The traditional British countryside, of hedgerows and small fields, of ancient farmhouses blending with the landscape, is changing.

28 What do you think is shown growing in figure 229? Look carefully at some of the nearer plants and you can see the fruit.

In fact, this is a new style of apple growing. The trees have only one branch, and the fruit is within easy picking distance. There are about 75 000 trees to the hectare in such a 'spindle tree' orchard. Contrast figure 229 with the traditional orchard in figure 219.

From the point of view of amenities many of these changes are unwelcome. Who is to decide whether increased efficiency in food production is more, or less, important than the beautiful countryside to which we are accustomed?

Figure 229

Have you eaten chicken lately, either at home or from one of the numerous street-side kiosks? If you have, it is almost certain that the chicken was bred indoors, without ever seeing the light of day. The same applies to the hens which lay most of the nation's eggs. This country now produces almost all its own poultry and eggs.

Some people consider that the methods of producing all this food are cruel and unnatural. On the other hand, it could be argued that life is not natural for any farm animals. And the hens are protected from bad weather, predators and illness. What do you think?

The farmer as a businessman

To help in the business side of farming, there are marketing groups like the Milk Marketing Board, and many crops are grown under contract. This means that, say, a vegetable canning factory, or a sugar refining firm, will enter a contract with a farmer to grow a certain weight of peas or sugar beet. Farms near the factories often specialise in growing such crops.

Even so, a great deal still goes to market. But often nowadays, the market is somewhere like the new Covent Garden market in London, rather than the local market town. Fast, modern communications have emphasised the position of London as the nation's leading food market centre, just as it is the chief centre for almost everything else.

Similarly, traditional methods still exist. Only about half our potato crop, for example, is harvested by machine. The descriptions you have just read of modern methods are trends, not firmly established ways of farming. The fact that they are trends, and therefore likely to continue in the future, is the reason why it is important to understand them.

Farming in Europe

Another very important influence on farming is that of international prices. Prices for farm produce are not entirely determined in this country, though our government has an influence on them. If the world price for certain goods is high, then so will be the price in the UK. Naturally, the reverse is also true.

International organisations have to be taken into account, particularly the EEC. This has a 'Common Agricultural Policy', which means that any changes in say, prices, have to be agreed by all the partners in the EEC which often causes considerable difficulty. The UK government has to try to follow the market's agricultural policy and at the same time get the best deal for the British food consumer (you!) and for our farmers. This is not an easy task.

International economics may not seem to be 'geography', yet it has a very powerful effect on the patterns of agriculture. Perhaps it can be illustrated by returning to the model with which this chapter started (page 145).

It was pointed out that high-speed transport made the model out of date. Nowhere in the UK was far enough from an urban centre to allow circular zones of produce to develop. But what about an international scale? If you substitute for the central town the industrial belt of Western Europe, what is the picture?

29 The industrial belt stretches from Liverpool/Manchester, through the West Midlands conurbation and London into northern France, Belgium, the Netherlands and into the Ruhr in West Germany. Look these places up in an atlas.

Figure 230 'Intensity of agriculture' in Europe

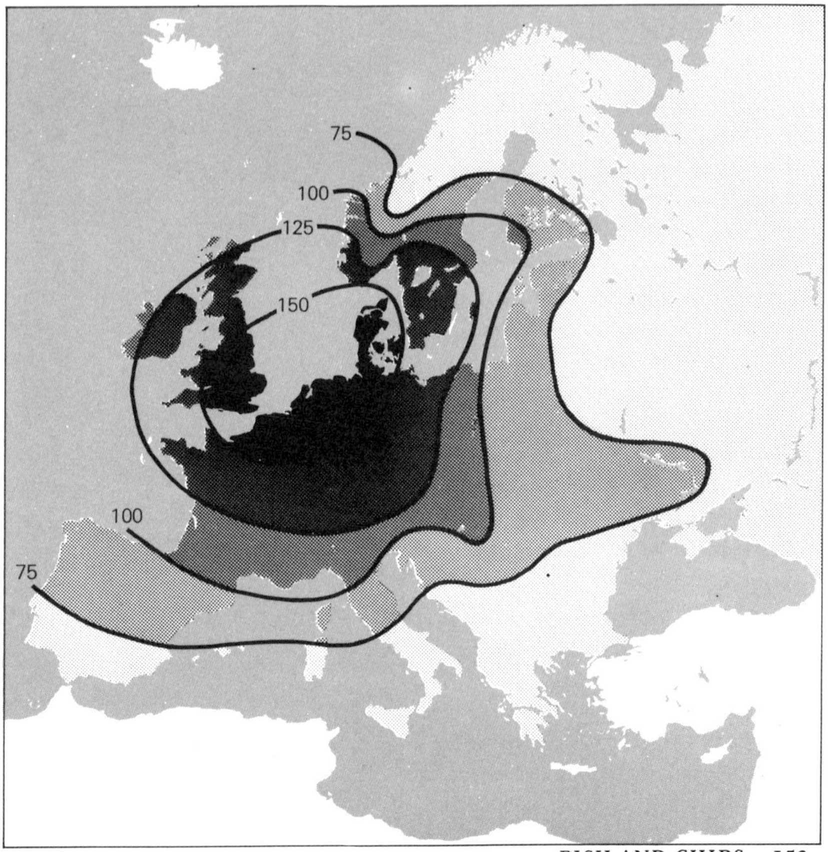

Figure 231

30 Look at the map in figure 230. 'Intensity' means the amount of time, money and effort that goes into cultivating a certain area of ground: the higher the number on the 'contour line', the higher the intensity. If you think of the industrial belt as a central 'town', can you see any similarity between this pattern and the model?

The fishing industry

There can be few people in Britain who have not been into one of the places shown in figure 231. The fishing authorities are keen to persuade people to eat more fish because it is becoming increasingly difficult for fishermen to make a living. Many different countries are competing for the fish, which means catching fish has to be as efficient as possible. In turn, this means that expensive boats and equipment are needed.

Fishing methods

Over half the British catch of fish is provided by trawlers. Trawling is a very obvious form of fishing. A net shaped like a bag is dragged through the shoal of fish and then hauled to the surface with the fish inside it. The mouth of the net is kept open by means of weights, floats and boards to ensure that the fish are trapped (figure 232 (a)).

The next most popular method is called 'seine netting'. For this type of fishing the boat sails round in a circle and lets out a length of rope, then the net, then more rope as it goes. Then the net is hauled in like a trawl (figure 232 (b)).

There are other types of net, including different sorts of seine. Can you find out about them?

At one time all fishermen had were their boat, their nets and their skill. Nowadays, all fishermen use scientific help. Many fishing boats have more navigation equipment than a cargo liner and as much underwater detection equipment as a warship. Even the smaller boats need some navigation gear and all boats use machinery for the hard work of hauling the nets.

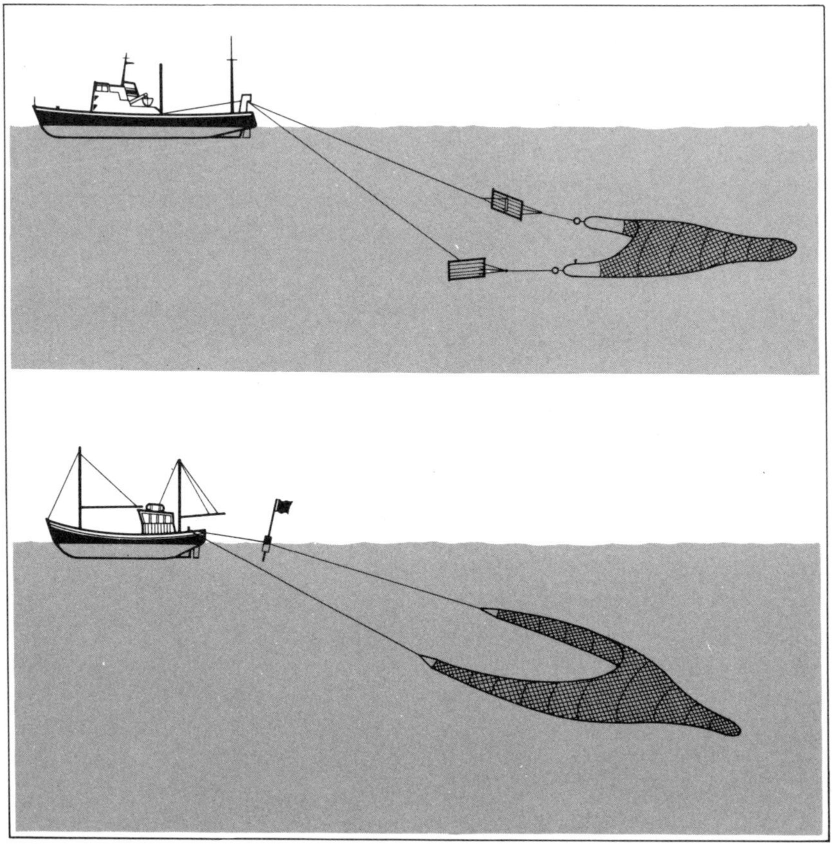

Figure 232 (a) Trawling (b) Seining

This adds up to considerable expense. Even the nets these days cost a lot. In 1973 a trawl net cost about £2 000. A special kind of seine net could cost up to £18 000!

It is hardly surprising that to be able to afford such costs, the fishing fleets are mainly owned by companies, rather than by individual men. To catch enough fish to pay for all the equipment, fishing has to be carried out further from home and often in poor conditions.

The fishing grounds

31 Look at the table in figure 233 and name the sea areas which form the 'distant', the 'middle' and the 'near' waters.

32 Now look up these sea areas in an atlas to see where they are.

33 What do you think is meant by a 'freezer' trawler?

All these fishing grounds are in some of the stormiest waters in the world, even the near water grounds.

The total catch of all fish in 1971 was divided as shown figure 234:

Figure 234

	£ thousand
Cod	42410
Other white fish	42123
Herring	4986
Total 'wet' fish	89519
Shellfish	6994
Total all fish	96513

34 You will notice that this table is given for the *value* of the fish caught. What percentage of the total is the value of herring caught?

35 What percentage is cod?

36 What kinds of fish are included under the heading 'shellfish'?

Fishing ports

Like other ports (page 137) fishing ports can be listed in the form of a league table. But how is the position in the league decided? Should it be by the value of the catch landed or by the weight? Perhaps the number of fishing boats operating from the port gives a better indication. But if so, should it be all boats, or are the bigger boats more important?

Figure 233 Fishing facts (white fish)

Type of vessel	Number of vessels (1972)	Regist. length	Areas fished and percentage of total supplies caught (British landings, 1971)	Main species caught
Freezer trawlers	37	73 m approx.	Iceland, Greenland, North Norway, Newfoundland, Labrador, Barents Sea, Bear Is. (9%)	Cod, haddock
Distant chilled fish trawlers	128	43 m and over	Same grounds as above, but with voyages lasting up to three weeks (24%)	Cod, haddock
Middle-water trawlers	173	33·5 m–42·9 m	Iceland, Faeroes, Rockall, Western Approaches, North Sea, English Channel, Irish Sea (29%)	Cod, haddock, whiting,
Near-water trawlers	152	24 m–33·4 m		Plaice, sole, turbot, and other flatfish
Inshore trawlers & seiners	815	12 m–23·9 m	North Sea, Irish Sea, and all coastal waters	Plaice, sole, cod, haddock
Other inshore vessels	893		All coastal waters (38%)	All white fish including shellfish and sprats

Figure 235 shows two tables of facts about UK fishing ports. There is a great deal of information in the table, so here are some questions to help you to sort it out.

37 Which ports come first, second and third for the total of all fish caught?

38 Which come first, second and third for white fish?

39 Are these the same as for herring?

40 Which port has the largest number of freezer trawlers?

41 Is this the same as the port with the largest number of trawlers in the table?

42 Write down which methods of measuring the importance of fishing ports are the most useful. Give your reasons.

43 Trace an outline of the UK from an atlas and mark on it the ports listed in figure 235. After the name of each port put in brackets its order of importance, using whichever method you have said is the best in your answer to question 42.

Selling the catch

The fisherman is the first link in a

				£ thousand
	Wet fish			
Place of landing	White Fish	Herring	Shellfish	Total all fish
England and Wales (fishing ports)				
North Shields	3247	174	132	3553
North Sunderland	65	2	52	119
Hartlepool	226	—	25	251
Scarborough	475	7	29	511
Whitehaven	183	189	112	484
Hull	18503	—	—	18503
Grimsby	21898	5	17	21920
Bridlington	388	—	32	420
Boston	1	—	95	96
Lowestoft	4765	5	—	4770
Great Yarmouth	5	5	5	15
Southend and Leigh-on-Sea	90	—	96	186
Brixham	289	1	16	306
Newlyn	456	—	41	497
Milford Haven	613	3	34	650
Burry Inlet	—	—	84	84
Fleetwood	5158	—	38	5196
Scotland				
Oban	521	253	309	1083
Ullapool	1673	817	487	2977
Lossiemouth	541	3	104	648
Buckie	1008	11	150	1169
Fraserburgh	1335	109	159	1603
Aberdeen	11016	13	14	11043
Leith	1418	2	221	1641

Figure 235 Fishing port facts (a) (table above) Fish landed, 1971. (Source: Ministry of Agriculture, Fisheries and Food) (b) (table below) Number of trawlers, 1973

Aberdeen		Granton		North Shields		Hull		
Fr.	Oth.	Fr.	Oth.	Fr.	Oth.	Fr.	Oth.	
0	117	0	17	0	8	36	62	
117		17		8		98		TOTAL

Grimsby		Lowestoft		Milford Haven		Fleetwood		
Fr.	Oth.	Fr.	Oth.	Fr.	Oth.	Fr.	Oth.	
9	103	0	88	0	16	1	42	
112		88		16		43		TOTAL

Fr. = Freezer trawlers. Oth. = Other trawlers.

chain which ends with you eating the fish. Once landed, the fish have to pass along the other links in the chain. At the major ports there are auctions where many of the fish are bought by wholesalers who then sell them to retailers at the big fish markets. Where do you think the biggest fish market is? In one of the ports, Hull perhaps? No. Like many other things mentioned in this book, it is in London, in this case at Billingsgate market.

Apart from this traditional way of buying fish, however, increasing quantities are bought 'on contract' by the big fish-processing companies who make food like fish fingers. Buying on contract means that the firms agree to buy a certain quantity of fish from the shipowners.

In all cases, the fish has to be transported quickly, before it goes bad, so that fast road and rail connections are important. Even here, however, things are changing. Fish frozen on the trawler and bought under contract by a fish processing company is not 'in such a hurry' because it will keep for much longer.

44 Knowing what you now do, you should be able to say from which sea areas the fish fingers in figure 236 are likely to have come. (Clue – frozen fish from freezer fishing boats.)

In one way or another, this chapter has been about some of the food you eat. Do you think that farming in this country should be allowed to expand so that we can grow more of our own food? Or should we produce more industrial goods and buy more food abroad? What about fishing? This provides a very nutritious food which, although it is landed at ports, is not 'imported' from another country. Should fishing be expanded? Have you ever heard of 'fish farming' – providing special

Figure 236

breeding and growing grounds for fish, in the same way that the farmer does for sheep or cows? In a world which is growing ever shorter of food, this might be the next breakthrough if a way could be found to do it on a large scale. It is already done on a small scale.

PROJECTS

1 Make a list of the food bought by your family during the course of one week. For each item, find out if it is manufactured and if so, what the ingredients are.

Then for each ingredient, or food item if not manufactured, find out where it came from. Try to work out roughly the percentage of your weekly shopping which was British in origin and how much came from abroad.

2 Write to: The Association of Agriculture, 78 Buckingham Gate, London SW1E 6PD for details of their farm study scheme, and for information about particular farms. Try to get information about different types of farms in different parts of the country.

3 It may be possible to arrange a visit to a farm, either near your school or during a fieldwork visit. If you can make such a visit, this chapter should help you to decide what questions to ask and what information to look for and record.

Conclusion

The first chapter of this book was about the way people 'see' things. In the rest of the book, you have seen how, because they see things differently, people disagree about many aspects of life in Britain. You have also been asked to think about the way decisions are made and how difficult it can be to make them when there is conflict about what should be done.

Conflict, however, is part of everyday life. We can never expect everyone to agree about everything. What we must do is to make sure that decisions are only reached after everyone's views have been heard. Also, everyone should be helped to understand as much as possible about the argument.

In Britain, many decisions which change the use of land are made by **planning authorities**. If the results affect the whole country, decisions will be made by the national government. Changes which affect a region will be decided on by **regional** planning authorities. Most local decisions are made by **county** planning authorities. Many of these last changes are small and some people think that local communities should have more say in these decisions.

Nowadays, many county planning authorities are asking for help from local people before they make any decisions. But even at the regional and national level, people can, and should play their part in influencing the planning authorities in the way they deal with the environment.

This book has tried to show you some of the processes at work in changing the environment. If you understand the processes, you should be able to play your part in helping to make decisions about the way *your* environment is treated. You can play some part even now, but when you leave school and have a family and can vote in elections, if you remember some of the things from this book, it could help you to be a very useful citizen of the UK.